Body Messages

Body
Messages

THE QUEST FOR THE PROTEINS OF
CELLULAR COMMUNICATION

Giamila Fantuzzi

Harvard University Press

Cambridge, Massachusetts · London, England   2016

First printing

Library of Congress Cataloging-in-Publication Data

Names: Fantuzzi, Giamila, author.
Title: Body messages : the quest for the proteins of cellular communication /
    Giamila Fantuzzi.
Description: Cambridge, Massachusetts : Harvard University Press, 2016. |
    Includes bibliographical references and index.
Identifiers: LCCN 2016009255 | ISBN 9780674088948
Subjects: LCSH: Cell interaction. | Cytology--History. | Cytokines. | Fat cells. |
Inflammation. | Metabolism.
Classification: LCC QH604.2 .F36 2016 | DDC 571.6--dc23
LC record available at http://lccn.loc.gov/2016009255

To Rosanna and Giulio, my parents,

who gave me life, love, and liberty

To Davide, my brother,

who taught me how to read and write

To Sergio, my beloved,

who gave me hope, heart, and happiness

Contents

Foreword

Dr. Hannah Landecker

HOW DOES ONE PART of a complex body know what another part is up to? How do cells and organs get it all together to react to an outside stimulus, incorporate food, fight infection, or incite growth? As an itinerant observer of science both past and present, my own dominant feeling about biology's answers to these questions is usually one of profound wonder that we are alive at all, a sense only augmented by the book you are about to read. Herein Giamila Fantuzzi describes *body messages*—proteins that move through the bloodstream between cells and tissues, harmonizing, conveying, and inciting bodily goings-on in an unceasing pullulation of activity, wondrous and intricate.

In a scene of popular science writing often dominated by genes and evolution or, alternatively, the biographies of particular diseases or famous individuals, it is illuminating to enter into this world of bodily intercommunication undergirding fundamental and cross-cutting processes of inflammation and metabolism, thereby understanding the complex posttranslational lives of proteins in action. One is struck by the paradoxically simultaneous wealth and dearth of biomedical knowledge about these proteinaceous actors. We know so much molecular detail about these entities and processes that still, to a great degree, manage to show up the vast acreage of what we don't grasp—

particularly when rheumatic or metabolic disease processes continue to evade our pharmaceutical control. There are a lot of human actors in the story here too, stories of intellectual passion, long hours at laboratory benches, and, difficult but necessary, the tensions that emerge when the object of science is at once scientifically interesting, medically important, and commercially valuable.

As a historian, I am struck by how far away from and yet how close we are to the first articulations of the "chemical messengers" of the body in the early twentieth century. This was the time when the first molecules understood as messengers between cells were chemically isolated from blood and tissues, and the term *hormone* was coined to name them. Before entering onto the more recent story of "body messages" that Fantuzzi lays out for us, it is worth briefly revisiting this earlier era. For this inception of the chemical body message concept— at somewhat more of a cultural distance than cytokines and adipokines—provides a number of useful thematic lenses through which to read *Body Messages*.

At the turn of the twentieth century physiologists were trying to figure out how the body "knows" about the food it is eating and reacts with processes and substances suitable to its digestion. Ivan Pavlov, who had famously conditioned dogs to salivate in response to a sound after having repeatedly presented food along with that sound, held that all such digestive secretions were nervous in origin. Experiments had shown that even down in the duodenum, the first part of the intestine directly beyond the stomach, pancreatic juices and bile flowed in, timed to meet food. Moreover, the digestive juices even seemed to be specific to the nature of the food, varying in composition whether the nutriment was mostly protein or mostly starch.

Yet the pancreas was at some distance from the intestine—how did this funny little organ know when and what to secrete with such temporal and material precision? Pavlov thought that the intestinal wall must be able to "taste" the food, and send the appropriate cue via the vagus nerve to the pancreas, stimulating it to action. Two British physiologists, William Bayliss and Ernest Starling, argued by contrast for the existence of a chemical "mechanism of internal correlation" between organs.[1] Experimenting in dogs, they showed that pancreatic secretions were stimulated even if all nervous connections between intestine and pancreas were cut. They isolated a substance from the duodenal epithelium that incited pancreatic secretions upon injection into another animal. This substance, which they called *secretin*, could be derived from dogs and evoke pancreatic secretions in frogs, rabbits, cats, or monkeys; in turn, secretin prepared from "the intestine of man, cat, monkey, rabbit, fowl, salmon, skate, frog, or tortoise" would excite pancreatic secretion in dogs.[2] Secretin was, they said, a *chemical messenger*, a transmission between the cells lining the intestine and the pancreas.

We might reflect that in Starling's London, six to twelve deliveries of mail could occur in a single day. In principle, this made possible the serial exchange of several letters between individuals at a distance. Life in the city was thick with messages and with messengers; one might receive or send a letter, or a telegram, or several of each. The relatively new technologies of telegraphy and telephony were demonstrating the possibilities of long-distance communication, changing the culture of space and time, and managing the coordination of actions across countries and oceans.[3] Just as a line needed to be laid for electric telephony but a mobile conveyance was needed to get a letter from one place to another, and one might happen in the moment and the other with slight delay, Starling distinguished between the nervous reflex and the chemical reflex, one being instant and the other taking some time, the former evolutionarily newer than the latter.

Perhaps it was the application of extracts of tortoise to control dog digestion that made Starling begin to think of these chemical entities as a new category of universal biological communicators, particular neither to humans nor to salmon, but instead a biological principle true across species. Only somewhat later, Starling sought a new term for his chemical messengers. Perhaps apocryphal, the story is that Starling dined with a Cambridge professor of classics, who suggested the Greek verb ορμαω, which meant to excite or arouse.[4] In 1905, the chemical messenger became a *hormone*, and we have used this term ever since. Coining the word was of course important, given its frequent use from then on, but often forgotten is the theoretical impulse that inspired it: to capture a theretofore unnamed and unknown phenomenon, a property of living things that Starling called the "chemical nexus" between organs. This chemical nexus was the sum total of the chemical messengers "speeding from cell to cell along the blood stream [coordinating] the activities and growth of different parts of the body."[5] It was distinct from nervous transmission, evolutionarily older, and involved the chemical transmission of "a message of some sort" from one set of cells to another.[6] And, according to Starling, it was the ultimate aim of medical science:

> If a mutual control, and therefore coordination, of the different functions of the body be largely determined by the production of definite chemical substances in the body, the discovery of the nature of these substances will enable us to interpose at any desired phase in these functions and so to acquire an absolute control over the workings of the human body. Such control is the goal of medical science.[7]

From this story, we can draw a number of frames to take to the more recent elaboration of body messages recounted in the book you are about to read. First, the question of *action at a distance* has been animating physiology for a long time—indeed, it still does. How does a body deal with the many inputs coming from the outside world, everything from nutrition to infection, and coordinate the spatial and temporal ballet of response? This question is all the more urgent thanks to the numerous ways in which coordination can fail. The interviews and stories of discovery unfolded by Fantuzzi in the book's chapters are fragments of maps of this remarkable dynamic spatio-temporal coordination of metabolic, immune, and growth processes. As she shows, the elaboration of new fragments often makes the overall picture of the territory shift. We might be appropriately humbled by reading Starling's optimistic prediction of 1905 of absolute control over the workings of the human body through knowledge of these substances.

Second, what was visible at the beginning was a general concept of a *chemical nexus* between disparate organs. Yet the triumphant rise of endocrinology in the early twentieth century and the apparent solidity of this category *hormone* itself contributed to the specialization of scientists and body parts and diseases into artificially separate immune, endocrine, and metabolic "systems" and disciplines, often making one organ and its troubles the province of the cardiologists and another the territory of the nephrologists. As Fantuzzi shows, it is useful to cast aside the apparent solidity of these conceptual categories—to remember that they are categories given by the history of science and not by nature—and think again through the nexus of body messages spanning the phenomena of adiposity and immunity, nutrition and infection.

A third frame for reading this book is suggested by the cultural context of Starling's chemical messengers. Inescapably, scientific questions, explanatory frameworks, and the narratives told about findings and therapies change along with the fashions and techniques of human communication. Culture shapes what we see, how we name it, and the bigger picture we fit the empirical detail into. If Marshall McLuhan had been a biologist, would he have worked on cytokines and adipokines? The body depicted by Fantuzzi and her interlocutors is the body proper to an age of burgeoning technical forms of information and communication.

Finally, thinking about Starling and his time, we might also observe that contemporary biologists should have dinner with classics professors more often. Perhaps then we would have a more felicitous and approachable vocabulary for the remarkable molecules that animate our lives and our sciences, whose stories are unfolded in the pages that follow. To Dr. Fantuzzi we owe gratitude for bringing these entities and their explanations out of the dense, acronym-thick domain of back issues of specialist journals and to a wider audience than they have had before.

Body Messages

Prologue

THIS IS A BOOK about the research process that led scientists to the discovery of a group of molecules that act as carriers of information among the cells of our body. Because investigators classify these molecular mediators under separate categories, I have chosen to refer collectively to them as *body messages*.

The upcoming chapters emphasize the variety of logical paths and experimental approaches that allowed identification and characterization of body messages. By presenting as many of the alternative strategies investigators engaged in when addressing a specific question, this book attempts to render the wonder of the unknown that is a researcher's main drive, while showing that the paths leading to each discovery were often all but linear, that the roads eventually chosen were not the only ones available.

In biomedical research, studying the physiology of the healthy state is so intertwined with investigating pathological conditions that the two become almost indistinguishable. Therefore, this book reviews the process of discovery and characterization of body messages by taking into account the role these molecules play in both physiology and pathology. These two sides of biomedical investigation cannot be separated without irreparably damaging them, blurring the distinction between basic and applied research. Indeed, as we will see, investigators identified body messages by starting both from the physiology of the homeostatic state and the pathology of dreadful illnesses. Along with the emphasis on the unreasonableness of distinguishing between basic and applied research, the book also stresses the importance of studying the body as a whole, of removing the artificial separations imposed by specialized disciplines, a theme that recurs throughout the upcoming chapters.

Linked with the process of body message discovery are the experiences of the scientists who carried out that process. In preparing this book, I interviewed twenty researchers—selected among those who contributed most significantly to the field—asking details about their discoveries and inquiring about their life and education. So interesting and copious were their answers that they could be the matter of a whole other book; nevertheless, I tried to incorporate the essential episodes of these scientists' lives into the narrative. Since many milestone anniversaries in the discovery of the body messages we are going to review have recently passed, published memoirs by the protagonists of the initial events have also greatly helped me piece together the puzzle. Along with scientists' recollections, I reconstructed the discovery process based on published findings, on reports of the original experimental results. However, as thorough as one wishes to be, mentioning each piece of evidence is an almost insurmountable task, with space unfortunately preventing inclusion of the crucial work of countless investigators.

After briefly introducing the author and the general context, Chapter 1 outlines basic aspects of biology as well as the concept of body messages and their receptors. This section also includes a brief review of inflammation and metabolism, the two fields of investigation that run throughout the book. This chapter serves as a primer for those who are not experts in the field to get acquainted with fundamental biological and physiological principles. Although readers already conversant with these topics may wish to skip these pages, several important concepts discussed throughout the book are first introduced here.

Chapter 2 familiarizes readers with ways of conceptualizing biomedical research, presenting various methodological approaches to scientific inquiry. These include observational and interventional studies, the principles of *in vitro, ex vivo,* and *in vivo* experimentation, the difference between asking what something can do versus what something is needed for, as well as the advantages and disadvantages of identifying a molecule beginning with its function or its structure. Most of the concepts introduced in this section will be helpful in framing and understanding the process of individual discoveries that constitute the gist of this book, though—as for the previous chapter—readers with expertise in these matters may choose to rapidly skim this section.

In Chapter 3 we enter the core of the book with the history of identification of members of the interleukin-1 family. On top of presenting the discovery of each member of this family, this discussion also serves as a review of the ways in which the process of body message discovery changed over time from a function-first to a structure-first approach, and to examine how this change affected investigation and understanding of molecular mediators.

The importance of balance between pro- and anti-inflammatory molecules is the topic of Chapter 4. Intestinal homeostasis is the setting for discussion of identification and characterization of four types of body messages—tumor necrosis factor, interleukin-10, chemokines, and annexin A1—that exert potent pro- and anti-inflammatory effects.

The protagonists of Chapter 5 are molecular decoys, mediators that inhibit the activity of body messages, some of which have been turned into powerful pharmacological treatments. Here we also talk about chromatography, a technique that has been essential in identification of molecular decoys and other body messages.

Interleukin-6 and the coordinated changes this cytokine induces in the liver—a process known as the acute-phase response—are the topic of Chapter 6, where the discourse begins to drift from inflammation to metabolism. In this chapter we encounter proteins like CRP, SAA and hepcidin while discussing the role of genetic diversity as a tool of discovery as well as the potential importance of biological uncertainty and ambiguity.

Chapter 7 deals with fat, comparing and contrasting the discovery and biology of leptin and adiponectin, two body messages that revolutionized the way we conceptualize the role of adipose tissue and that bring us all the way into regulation of metabolic pathways.

Finally, a short epilogue reiterates and rounds up several themes that course throughout the whole book, including the variety of conceptual and technical approaches to body message identification as well as the futility of neatly separating basic from applied research and of sharply delineating fields of investigation.

As this outline indicates, body messages are the book's main characters. Among the thousands of molecules that act in such fashion, I selected those that have been part of my own research, which has mostly focused on the interconnected fields of inflammation and metabolism. Much of the physiology of a healthy organism and of the pathology of a diseased one can be interpreted within the framework of these two general processes, thus placing inflammation and metabolism at the very core of biomedical research.

Though the book's main theme is the process of body message discovery, considerable space is nevertheless given to the biology of these mediators and the consequences of their identification for medical practice. While attempting to be as current as possible, I specifically shied away from last-minute knowledge, unless absolutely necessary, since trying to include up-to-date information in a book like this would be a foolish task. At the pace new discoveries are made, even the leanest specialized journals are unable to keep up. Instead, identification of each mediator is framed with a background that should allow readers to understand the relevant biological pathways and appreciate the implications, both practical and theoretical, of identification and characterization of these body messages. Wherever appropriate, involvement of the biotechnology and pharmaceutical industry is presented, together with a few historical notes and my own reflections. The book includes plenty of references for those who wish to explore any of these issues in more depth.

1

The Medium Is the Message, the Message Is the Change

I WAS BORN IN Milano, Italy, in 1964, the same year Marshall McLuhan published his book *Understanding Media: The Extensions of Man*, which contains the famous sentence "The medium is the message."[1]

Milano was a major industrial, financial, and cultural center at that time, having rebuilt itself after major bombings during World War II. I grew up in one of the neighborhoods erected at the extreme borders of the city to accommodate the scores of workers flocking to its factories: huge apartment buildings, few stores, limited public transit. But what was lacking in infrastructure was made up by strong solidarity among the working-class people who had relocated there. And we kids loved the open spaces and construction material seemingly left over for our enjoyment.

By the mid-1980s, Milano had undergone a major shift from a manufactory-based to a postindustrial economy. Fashion, design, advertising, and public relations were the new mantra: communication and information became everything that mattered, or so it seemed. An inane cult of the image and of easy money eventually devolved into an era of corruption scandals. At the time, however, young people like me wanted to be trendy: communication—not substance—was the vogue.

I was studying biology at the University of Milano during those years: how could I not decide to focus on the body's information network? I vividly remember thinking I wanted to study the nervous, endocrine, or immune system, the three major channels of communication in our body. I wanted to understand biological information systems, not having the least interest in anything that had to do with structure. I guess I was a child of my time.

Molecules, Cells, Tissues, Organs, and Systems

Our body is made of cells, which form tissues, organs, and systems. Stomach, lungs, and kidneys are examples of organs, composed of different types of tissues organized into structures that accomplish specific functions: digesting food, breathing, filtering blood. The tridimensional geography of an organ, the way in which cells are organized into physical structures such as the alveoli in the lungs (where exchange of air takes place) and the glomeruli in the kidney (where blood is filtered), is essential in determining function.

But that's not all there is to it. If tissues are made of cells, cells are made of molecules. These can be simple small molecules like water as well as big and complex ones, such as proteins. Not only are cells made of molecules, cells make molecules. They do this to replace damaged parts but also to carry out specialized tasks. For example, cells in the wall of the stomach make hydrochloric acid that is required for digestion of food. When these cells stop making acid (which can happen for a variety of reasons), the stomach is unable to perform its function even though its structure is not initially altered. Thus, an organ cannot work properly unless both its geography and the set of molecules produced by its cells are safe and sound.

Organs that perform related functions are considered part of a system, such as the digestive, respiratory, and urinary systems. The word *system* also refers to those parts that accomplish a common task but are not located in a single place, being instead distributed throughout the body. These systems are in charge of dissemination and communication, both with the outside world and within our internal space. The skeletal system—muscles, bones, and joints—lets us move and speak, sustaining communication with the environment and with other people. The nervous system not only conveys messages from the outside and coordinates our interactions with the world, but also transmits, organizes, and integrates information about the state of every part of our body. By making blood reach every nook in our organism, the circulatory system—heart and vessels—delivers oxygen and nutrients to each cell and collects waste for disposal. Another vital function of the circulatory system and of blood, one that is specifically relevant to our topic, is to distribute molecular messengers produced by cells, including those of the endocrine and immune systems.

Whereas the skeletal, nervous, and circulatory systems are relatively easy to define and circumscribe, classifying components of the endocrine and immune systems is instead more difficult, particularly as our understanding of these systems evolves. Scientists traditionally defined the endocrine system as a series of glands, which are structures made of cells whose main job is to produce molecules called hormones and to release them into the blood, which carries them away to various locations. The thyroid and parts of the pancreas are examples of endocrine glands, respectively producing the hormones thyroxin and insulin. These hormones act as body messages but they are not the main characters of our stories.

With time, however, the definition of endocrine system became muddled, since cells that are part of many tissues, not just those present in endocrine glands proper, are now known to produce molecules that act as hormones, making strict classifications quite difficult. An example of a mediator that is hard to classify is the body message hepcidin (discussed in Chapter 6), a molecule with hormonal function made by cells of the liver, an organ not usually considered part of the endocrine system.

Things get even more complicated when it comes to the immune system, guardian of the integrity of our cells and tissues, on the lookout for any kind of danger, be it an infectious agent, a mutated cell that has become cancerous, or a tissue that might have been damaged for a myriad of reasons. We can easily categorize certain parts of the organism as belonging to the immune system: thymus, spleen, tonsils, lymph nodes, white blood cells. But what about the wall of our intestines, which contains the largest collection of immune sites in the whole body? How should we classify it? Is the intestine part of the digestive system or the immune system? And where should we place the body message interleukin-6 (discussed in Chapter 6), mostly produced by immune cells but often acting as a hormone that profoundly affects other organs, including the liver? And the liver itself? Oh, the liver, what should we make of it? Categories, divisions, and nomenclature reflect the state of the knowledge at any given time, changing and evolving with our interpretation of the body's workings.

Body Messages, Their Receptors, and Signal Transduction

What I decided to refer to as body messages are molecular mediators, carriers of information from cell to cell, molecules that act as hormones, neurotransmitters, regulators of inflammation, metabolism, and other functions. Irrespective of how scientists decide to categorize them, body messages share common features.

A first characteristic of body messages is that they are molecules made by cells. They can be any type of molecule. Tiny two-atom nitric oxide (NO, one atom of nitrogen and one atom of oxygen), a crucial regulator of blood pressure. Steroids, like the sex hormones estrogen and testosterone and the stress hormone cortisol, made by chemical modifications of the skeleton of cholesterol. Lipids, such as the prostaglandins we often choose to block with drugs like aspirin to achieve relief from fever, pain, and other unpleasant symptoms.[2]

Despite the extreme diversity in the molecular structure of body messages, the mediators featured in these pages all take the form of proteins. This means that a gene—a specific sequence of the letters (nucleotides) ATCG in our DNA—contains the code for each of the body messages we will review. To make a protein, the nucleotide sequence in the corresponding gene gets transcribed from DNA into messenger RNA and eventually translated from the code of nucleic acids (DNA and RNA) into that of amino acids, the building blocks of proteins.

It is often said that humans share about 99.9 percent of their DNA sequence. What this means in practice is that if I take all the DNA from my cells (or yours) and read it letter by letter, nucleotide by nucleotide, the order of ATCGs would be identical, on average, to that of every other human being on the planet, with the exception of 0.1 percent of the entire sequence. That 0.1 percent of DNA, together with—one should never forget—the specific events and circumstances of our lives, is what makes each of us a unique individual. Since the body messages we will discuss are proteins, which have their corresponding genes, some of that 0.1 percent variability affects them in multiple ways.[3]

We said that body messages are molecules made by cells. Any kind of cell can manufacture body messages. Some, like the hormones insulin and thyroxin, are only synthesized by a single specialized type of cell that appears to do little else other than pump out hormones whenever they are needed. Other body messages, such as the adipokines leptin and adiponectin (discussed in Chapter 7), are also made by a single type of cell (fat cells, or adipocytes, in this case) but these cells have many other important tasks to accomplish in addition to producing body messages. For example, on top of making adipokines adipocytes have the critical role of storing energy in time of nutritional excess and releasing it when needed. In many other cases, body messages are produced by numerous kinds of cells that perform a variety of additional jobs. Cytokines and chemokines, for instance, are made by various sorts of white blood cells—the main effectors of immune responses—but also by the endothelium, those flat cells that line the inside of our blood vessels, as well as by many other kinds of cells in multiple tissues and organs.

A second important property of body messages is that they need to move outside the cell in which they were assembled in order to convey their information.[4] Once factory cells have released it, the mediator can enter the blood and be carried away to deliver its message to distant places, in what is called an endocrine pattern of communication. Hormones like insulin and thyroxin, but also many other body messages, function in this manner, using blood and the circulatory system to reach their final destination. Other body messages instead carry their information only a few steps away from the cell that manufactured them, a pattern named paracrine. In this case, only the cells and tissues that surround the source of the message get the news. Messages that act in this manner are typically very powerful and can be highly damaging if they reach the blood in high concentration. Information about the presence of inflammation is often conveyed using the paracrine system, thus being for the most part confined to the site of damage or danger. The reaction to a mosquito bite is a clear example of paracrine communication: redness and itching are limited to the immediate vicinity of the bite. Finally, the most restricted type of news delivery is that of the autocrine response, in which the message acts only on the same cell that produced it, somewhat analogous to writing a memo to oneself. However, this is typically coupled with the response of nearby cells, as if the memo were attached to the refrigerator door so that other people in the household could act on it. Therefore, autocrine and paracrine responses often occur at the same time.

A third characteristic of body messages is that they deliver their information by attaching to specific structures on target cells, the receptors.[5] The analogy of lock and key is always called upon to explain the connection between a ligand, the body message acting as the key, and its receptor, the lock. The two molecules, message and receptor, have complementary structures that allow them to interact in a very intimate manner.

Receptors can be grouped in various families, but typically one end of the receptor sticks out of the enveloping membrane of the receiver cell, with the other end dangling on the inside of the cell, the cytoplasm.[6] When the message attaches to the receptor on the segment that sticks outside of the cell, this binding results in subtle changes to the structure of the receptor itself, transmitting modifications to the part of the receptor that hangs inside the cytoplasm. This modification, or activation, of the receptor sets in motion a cascade of biochemical events inside the cell, called signal transduction, that involves a series of intermediate molecules that eventually modify the behavior of the responder cell. Several kinds of messages, thousands of them, activate their specific receptors on a single cell at any given time: the process of signal transduction coordinates the information these messages convey, with the target cell behaving in a manner appropriate to the totality of the broadcast.

The Message Is the Change

Marshall McLuhan's statement that the medium is the message entered popular culture to such an extent that asking the question

"How do hormones carry messages?" on Answers.com gets you the following response, "Hormones act on cells that have specific receptors within the cells that respond to them. So in this case the medium is the message."[7]

In *Understanding Media: The Extensions of Man*, McLuhan argued that technologies profoundly change individuals and society irrespective of their content. Specifically discussing the meaning of his now famous sentence, McLuhan wrote, "For the 'message' of any medium of technology is the change of scale or pace or pattern that it introduces into human affairs."[8] Translating this concept into the world of body messages, one can say that the message of any biomolecular medium is the change of scale or pace or pattern that it introduces into receiver cells. Hence, the main message of the medium "insulin" is the ability of sugar to flow inside a cell that needs it as nourishment, a change in the scale and pace at which the receiver cell takes up sugar, and thus in the pattern of sugar distribution throughout the body.

As recognized by McLuhan himself, the economist Kenneth Boulding had already used this concept in his 1956 book, *The Image: Knowledge in Life and Society*. Here, Boulding defined the meaning of a message as the change that message produces.[9] The notion that the meaning is the change is essential not only to appreciate the complex biology of body messages but also to understand the history of their discovery. In fact, as we will see, it was scientists who were trying to uncover the mediators responsible for specific biological changes in the systems they were studying who identified the vast majority of body messages. For example, researchers discovered the first members of the interleukin-1 family, the protagonists of Chapter 3, while trying to find the molecules that cause fever, that is, a change in body temperature, and those at the origin of damage to bones and joints, which is a change in the skeletal system. As technology evolved, the process of discovery at times turned upside down: newer techniques made it easier to first discover a molecule, to identify the structure of the message, and only at that point try to figure out what the content of the message might be, to discern which changes that particular molecule would induce. Thus, as scientists invented new tools for biomolecular investigation, the discovery process partly changed from a function-to-structure direction to its reverse, a road from structure to function.

However, as will become obvious as we review the history of identification of each body message, beginning with the change remained the most widely used strategy throughout the whole history of message discovery, up until the present time, testifying to the power of this approach.

Two Topics

The body messages featured in this book are those I encountered at various stages of my own research. Although admittedly a biased choice, this set of molecules is sufficiently large and varied to tackle several critical issues concerning the general principles underlying the research process that led to identification and characterization of most body messages.

In the upcoming chapters, we will encounter mediators discovered as part of two separate disciplines, which correspond to the two topics of my research, inflammation and metabolism.

Inflammation

Inflammation is the body's first response to danger and damage. These may come from the outside in the form of infectious, physical, or chemical agents, or from the inside, such as trauma or cancer. Danger from the outside often ends up damaging tissues, thus adding an internal source of warning.

Cells perceive danger signals through a family of receptors called pattern recognition receptors—PRR in short—the discovery of which won Jules Hoffmann and Bruce Beutler the Nobel Prize in Physiology or Medicine in 2011.[10] Once PRR are activated by signals of danger and damage, intracellular signal transduction cascades lead to production of a plethora of mediators and to movement and activation of white blood cells—also called leukocytes—and other cell types, resulting in development of an inflammatory response.

Leukocytes come in different variants. There are monocytes/macrophages, which eat up bacteria, particles of all sorts, dead tissue, and anything else that should not be where it is. These cells also act as sentinels: once they sense danger through PRR, they rapidly get activated and send out countless mediators that start the inflammatory process. At a later time, the same cells produce a different set of molecules that help in terminating the inflammatory response and repairing any damage that might have occurred. Monocytes/macrophages are thus necessary to begin, shape, and wrap up an inflammatory response. Inappropriate functioning of these cells—either too little or too much—at any step of the process can be harmful. In case you are wondering about these cells' double name, they are called monocytes when present in blood, macrophages when in tissues.

Neutrophils are another type of leukocytes. These cells are present in large quantities in the blood and are called in at sites of danger by messages sent out by the sentinels and by other tissue cells. Once in place, neutrophils rapidly get to work, destroying microbes and producing mediators that help clear danger and damage but that can, in turn, be directly noxious to the body. Once neutrophils have finished their job, they pass away and are eaten up by macrophages.

Lymphocytes are the third major type of leukocyte. These small cells are typically associated with long-term immunity to individual infections, such as the protection afforded by vaccines. Lymphocytes are, in fact, the cells that precisely recognize specific molecules—called antigens—and react only when such molecules are around. One type of lymphocytes, B cells, produces antibodies, proteins that float around and stick to their respective antigens, alternatively inactivating them or flagging them for removal by other immune cells. The second type of lymphocytes, T cells, either directly kills cells that contain specific antigens or help the rest of the immune system work in a coordinated and efficient fashion.[11]

Scientists have traditionally imposed a sharp dichotomy on the immune system, with monocytes/macrophages and neutrophils considered part of innate (also called unspecific) immunity, the branch that quickly reacts against large classes of stimuli with generation of an inflammatory response. Immunologists have instead historically classified lymphocytes as the main protagonists of acquired (or specific) immunity, a slower but more precise and targeted response compared to the innate wing. Typically, innate immunity is considered necessary to activate acquired immunity, which then provides long-term protection against infections but can also be at the origin of allergic and autoimmune diseases or transplant rejections. Just like many other neat distinctions, this dichotomy has been extremely helpful in framing our understanding of immune responses; it has been relatively easy to reflect the biology of leukocytes into our human mirror of the concepts of innate and acquired immunity. However, experimental evidence eventually ended up smudging the tidy lines that separated the two arms of immunology. As a result, albeit still very much part of the teaching, researching, and rationalizing of the immune system, the barrier between innate and acquired immunity is being dismantled one brick at a time.[12]

Inflammation is an essential component of innate immune responses, fundamental to clearing up danger and damage of either external or internal origin. Limited, controlled inflammation is a beneficial and indispensable response. But when inflammation becomes chronic or excessive it turns into a harmful situation that needs to be dealt with. In the upcoming chapters, we will see how research on body messages shaped knowledge of the mechanisms underlying many acute and chronic inflammatory diseases, resulting—among other achievements—in development of widely used drugs.

Metabolism

The word *metabolism* derives from the Greek term for "change" and refers to the set of chemical transformations that allow cells and organisms to sustain themselves, grow, reproduce, and survive. Topics such as absorption of nutrients from food and their distribution around the body, conversion of one type of energy into another, as well as use of different energy sources are all covered under the general umbrella of metabolic research. Scientists who study metabolism also examine the chains of chemical interconversions in the cell, evaluate the tissues—fat, muscle, liver—that store, use, and control such energy, and investigate the hormones that regulate energy intake, distribution, and utilization. Metabolism has always been strongly interconnected with endocrinology, the hormones produced by pancreas (insulin and glucagon), thyroid (thyroxin), and adrenal glands (steroids) playing vital roles in the control of metabolic pathways. Indeed, the fields of endocrinology and metabolism are linked so tightly that, in most universities, they are brought together into a single department.

However, though we may associate the hormones of endocrinology—and thus metabolism—with the study of body messages, the concept of hormones' involvement in the exchange of biological information is relatively new. In her article "Postindustrial Metabolism: Fat Knowledge," Dr. Hannah Landecker elaborated on how our interpretation of metabolism has changed in concert with the transformations undergoing in our societies. Explaining how both the term and the concept of metabolism were established in the industrial era of the nineteenth century and initially interpreted within the framework of industry and factory work, her article discusses how—as societies changed from manufactory-based to information-centered environments (at least those of the Westernized world where biomedical research mostly takes place)—interpretation of metabolism was reshaped using the metaphors of regulatory and information systems rather than those of factories and industry. It was indeed research on metabolism that, in the first half of the 1970s (the years in which transition to a postindustrial economy became more prominent and visible in the richer countries), conceptualized the ideas of biological symbolism and intercellular communication by exchange of information, particularly as a result of analysis by the late Gordon Tomkins. "If we live in an information age, why wouldn't our biologies?" Dr. Landecker pointedly asks.[13]

Blurring the Lines

Just as a neat line cannot separate the two arms of the immune system, a sharp distinction between inflammation and metabolism no longer makes much sense, if it ever did. In fact, Dr. Landecker's description of the current interpretation of metabolism as "a zone of regulation, a space for the integration and coordination of external and internal signals, whose timing is as important as their nature" impeccably defines our current understanding of inflammation as well.[14]

If the same definition applies to two supposedly distinct areas of investigation, shouldn't one begin to wonder whether a separation between the two fields is at all warranted? Indeed, inflammation and metabolism are becoming more and more commingled as researchers study how metabolic pathways shape inflammation, while inflammatory responses are being extensively investigated as modulators of metabolism. This is not to say that an underlying current of research linking the two areas hasn't always been present. However, the rising prevalence of chronic diseases with both metabolic and inflammatory components—diabetes, cardiovascular disorders, arthritis, cancer, and many more—boosted and motivated scientists to investigate inflammation and metabolism as a single entity, rather than as separate processes.[15]

The body messages discussed in this book, though initially discovered and for the most part studied separately as part of either inflammation or metabolism, are an excellent example of the impracticality and impossibility of segregating these two fields from each other.

Small proteins known as cytokines are the characters of many of the chapters ahead. Cytokines' nomenclature can be daunting. Some cytokines are called interleukins, with a number designating the order in which scientists discovered them. Other cytokines, such as tumor necrosis factor, retain the name investigators assigned to them based on the original context in which they were discovered, even though some of this nomenclature has since lost meaning with the evolving understanding of these mediators' main biological functions. Yet other molecules that may qualify as cytokines, such as annexin A1, are not classified as such for historical and structural reasons, although their function overlaps in many ways with that of cytokines.

It was scientists studying inflammatory and immune responses who discovered the majority of cytokines. However, because inflammation brings about changes in metabolic pathways, one can as easily approach the very same mediators for their metabolic, rather than inflammatory, effects. Indeed, metabolic changes occurring in the context of infections and inflammation—fever, depletion of fat, alterations in liver function—have been at the origin of identification of more than one cytokine, as we will soon find out.

Some of the body messages we will encounter in the upcoming chapters are called adipokines because they are mainly produced by adipocytes, the cells that store fat. Researchers studying metabolism discovered adipokines and, to be sure, these molecules are involved in a major way in metabolic regulation. However, adipokines also play crucial roles in modulation of inflammatory and immune responses, again an example of overlap between scientific areas and of the challenging, at times pointless, attempts at forcing biology into human cultural categories.

2

Doing Research: Ways to Study Body Messages

BIOMEDICAL RESEARCH has a vast scope. At one end of the spectrum are the scientists who study the invisible world of molecules, trying to understand the precise shape of a protein or determine the function of a stretch of DNA, for example. At the other end are those who investigate whole populations, perhaps assessing the impact of a new vaccine or evaluating the risk of an epidemic. Between these two extremes stand the researchers who study health and disease by examining cells and individual bodies. Despite encompassing an extensive array of topics and approaches, general categories can be used to conceptualize research in biomedicine.

Observing and Intervening

One set of categories splits biomedical research into observational and interventional studies, terms with a precise definition in epidemiology but here applied in a more general sense.[1]

In an observational study, a scientist selects one or more objects of interest and investigates them without attempting to induce changes in the system. Indeed, the whole goal of observational studies is to avoid perturbations as much as possible, trying instead to act as an external impartial observer. However, the researcher should always be mindful that the act of observing ends up influencing the phenomenon under observation. Even the mere act of specifying the conditions under which an observation is to be carried out will inevitably influence the outcome.

Examples of observational research in the realm of body messages include a crystallographer who wants to identify the tridimensional shape of a protein, a molecular biologist who tries to understand the structure of a gene, or an immunologist who plans to quantify the amount of a molecule in the blood of patients with an autoimmune disease and compare it with the amount present in a group of healthy subjects. In this last case, deciding whom to include in each group— which patients and which healthy controls—is essential. What parameters should the researcher use to define the presence and absence of the specific autoimmune disease? Will the study include patients with mild disease, only those with the most severe form, or both? Do the researchers plan to study only individuals who have just been diagnosed with the autoimmune condition or are they instead interested in patients with long-standing autoimmunity? Which criteria will determine matching between the group of patients and the group of healthy control subjects other than presence or absence of the specific disease?[2] How and when will the phlebotomist collect blood samples to measure the molecule of interest? Which technique will the analyst choose for quantification? Which tests is the statistician going to employ to analyze the results? These and many other critical questions, including the essential decision about how many subjects the investigators should include in each group, will determine the quality of the data obtained while also influencing the shape of the results, that is, whether a significant difference in the amount of the molecule of interest will be detected between the two groups. These issues are not unique to investigations that involve human subjects. Researchers performing other types of observational studies need to ask and answer a comparable and no less daunting set of questions meant to maintain as much objectivity as possible in the interpretation process.[3]

In contrast to the passive approach of observational research, interventional studies aim to actively modify the system under investigation, perturbing it in specific, controlled ways.[4] Because the goal of an interventional study is to understand how the object responds to changes in its surroundings, here the scientist does not take the role of an external observer but rather that of a demiurge who shapes and manipulates the environment, pushing it to reveal itself.

Types of interventional research relevant to the science of body messages include, for example, a cell biologist testing whether cells in a dish grow faster when a specific molecule is added, a physiologist evaluating whether rats injected with a particular compound change how much food they consume, or an allergologist checking whether symptoms improve in a group of patients receiving a new drug. Although the scientist here has an active role in deciding how to modify the environment, once the intervention has been implemented the researcher needs to step back and become an observer again to measure the effects of the perturbations introduced into the system. Both before and after the intervention, investigators also need to address the same set of questions mentioned above for observational research, including a clear definition of the groups and conditions for the study, techniques, analytical methods, and so on.

Thus, scientists doing both observational and interventional research need to adhere to a set of equally rigorous rules in each phase of the investigation, including study design, implementation, analysis, and interpretation.

Tubes and Bodies

Another way of categorizing variants of biomedical research is to use the concepts of *in vitro, ex vivo,* and *in vivo* studies.

The term *in vitro* literally means "in glass" (from the Latin) and refers to the test tubes, dishes, flasks, or other suitable containers used to perform experiments, containers that were once made of glass and nowadays mostly of plastic. The *in vitro* concept can be applied widely, from the study of isolated molecules to that of cells that are no longer part of an organism but are instead kept alive in the laboratory, fed a liquid—called medium—that provides the necessary nutrients for survival, growth, and proliferation.

In vitro studies of the molecules that make up our bodies, including proteins and nucleic acids, fall under the purview of biochemical analyses.

These include methods to separate a specific molecule from a mixture of compounds in order to determine its characteristics, evaluation of interactions between two molecules (think of a body message attaching to its receptor, for instance), determination of DNA sequences, quantification of the amount of a given protein, and countless more.

A separate category of *in vitro* experiments utilizes cells grown outside the body. These can be cells freshly obtained from a person or an experimental animal. Leukocytes isolated from blood through venipuncture are among the easiest types of cells to collect from humans. Other human cell types can be accessed via biopsies or surgical specimens, while experimental animals allow investigators to retrieve and culture fresh cells from any tissue or organ.[5]

Use of freshly obtained cells—called primary cells—is extremely powerful, as it puts the researcher as close as possible to the body being investigated without significantly endangering the individual (when human subjects are involved).[6] There are limits to this approach, however, since only certain types of cells can be obtained without invasive procedures, while the quantity of cells that can be safely collected is often limited. Crucially, most primary cells can only survive and replicate for a few days once they have been removed from the body, thus restricting the type and extent of experiments scientists can perform.

To get around these issues, *in vitro* experiments often make use of cell lines. These are cells that were once obtained from a living being—a mouse, a rat, a human—and have been selected or manipulated to allow them to grow in an almost unlimited fashion in the laboratory, generation after generation. To make cell lines, scientists often exploit the ability of cancer cells to divide and reproduce without limits, one of the central characteristics that make cancer such a deadly disease. Thus, many of the cells that are now cell lines were once part of the tumor of a person living (and often dying) years ago, decades even. The most famous cell line, called HeLa, consists of cells that were once part of the cervical cancer that in 1951 killed Henrietta Lacks, an African American woman from Virginia whose story Rebecca Skloot magisterially narrated in her book *The Immortal Life of Henrietta Lacks.*[7] There are literally thousands of cell lines derived from all sorts of tumors: a cancer of the cervix in the case of HeLa cells, cancers of the colon, liver, lung, or brain in lines with names like Caco-2, HepG2,

A549, and U118-MG. Other cell lines do not come from a tumor but from an embryo or fetus, that stage of life in which cell proliferation proceeds at full speed without need for malignant transformation. Still others derive from normal healthy cells that have been purposefully infected with viruses or otherwise manipulated after collection to allow them to replicate extensively *in vitro*.

Specialized companies maintain these cell lines and sell them to scientists: access the online catalog, select the product and put it in your virtual cart, pay a few hundred dollars with your academic credit card, and in a matter of days you will receive a vial of frozen cells. You will carefully thaw them following a well-defined protocol and put them in a dish inside a warm and humid incubator that mimics the body's inside. *Et voilà*, you now have in your laboratory cells that once belonged to somebody, somewhere.[8]

Now that you have cells growing in a dish, be they a cell line or freshly obtained primary cells, what do you do with them? You can use them to investigate all sorts of questions, from evaluating which molecules they produce, perhaps in a purely observational-type study, to using an interventional design to actively manipulate their environment in order to check how they respond to changes.[9]

To recap, *in vitro* experiments use molecules or cells outside of their bodily context, placing these research objects in an artificial environment that allows scientists to observe, manipulate, and analyze them in ways that would not be otherwise accessible had they remained part of the whole organism.

The concept of *ex vivo* research partly overlaps with the *in vitro* approach. Again from the Latin, *ex vivo* means "from the living," referring to cells, tissues, or organs removed from a body and studied outside the confines of the intact organism. A biopsy of the wall of the intestine a gastroenterologist collects during a colonoscopy and an investigator maintains in a dish for a day or two is an example of an *ex vivo* setting, as are organs freshly harvested from experimental animals and artificially maintained alive outside that animal's body. Because every cell, tissue, and organ and many biomolecules studied *in vitro* were once part of a living organism, one of the most important differences between the *in vitro* and *ex vivo* approaches is the time elapsed between removal from the body and experimental study, a time that can be extremely long for *in vitro* experiments, but is always necessarily short for *ex vivo* ones. Sometimes the two terms are interchangeable. For instance, leukocytes isolated from an individual's blood and cultured in a dish can be referred to as either an *in vitro* or an *ex vivo* setting depending on the subtleties of the experimental design.

At variance with the *in vitro* and *ex vivo* approaches that use parts of a body (molecules, cells, tissues, organs) detached from the organism to which they once belonged, scientists performing *in vivo* studies observe or intervene directly in living bodies that can be either experimental animals or human beings.

Investigators use experimental animals to perform manipulations and interventions that would be ethically and practically impossible in humans. While biomedical scientists make use of all types of animals, from nonhuman primates to fish, flies, and worms, it is rodents, particularly mice, that fill most of the rooms of today's animal research facilities. Mice are mammals, like humans, but they are small (a couple can fit on the palm of your hand) and reproduce quickly (pregnancy lasts less than a month), making them relatively easy to handle. In the last four decades, development of genetic engineering techniques contributed to sharply increasing the use of mice as experimental subjects of biomedical research, a development whose importance the Nobel committee recognized in 2007 by awarding the Nobel Prize in Physiology or Medicine to Drs. Mario Capecchi, Martin Evans, and Oliver Smithies for their leading role in establishing the field of manipulation of the mouse genome.

Researchers often use experimental animals to model human diseases, with the ultimate goal of understanding the origin and mechanisms of pathologies that afflict humankind, in the hope of developing strategies to prevent, treat, or cure such illnesses. Most investigators working in the area of disease modeling try their best to establish and use experimental systems that mimic as much as possible the corresponding human condition, so that the results obtained can be relevant and translatable to the medical setting. However, due to differences between the physiology of laboratory animals and that of humans as well as countless practical issues that range from cost to duration of the experiments, modeling human disease in experimental animals inevitably results in an approximation. This is a particularly troublesome issue for pathologies that have complex causes and those that typically develop over the course of decades—think of diseases like cancer, Alzheimer's, or hypertension. Due to these and additional concerns, scientists who study human diseases in experimental animals are always walking a fine line between using the available models, even if imperfect, and trying to establish new ones that might be more appropriate, if more cumbersome. Debate on these topics is one that never ceases both inside and outside the scientific community, and rightly so.

Like any other type of research, *in vivo* studies in humans—called clinical studies—can take the form of observation or intervention. Although in biomedical research observing can mean analyzing trends without requirement for an individual's direct participation (examining the connection between a source of contamination and incidence of cancer in a particular geographical area, for instance), an observational clinical study requires a person's active participation as the subject of scientific research. In this situation (say, observation of whether a newly discovered molecule is present in people with a certain disease), it is paramount that the individual is fully informed about the details of the clinical study and willfully consents to participate. Even though the term *observational* may seem innocuous, the potential for harm should never be discounted.[10]

In vivo observational studies in humans can help answer countless questions. Relevant to the study of body messages are measurements of mediators in blood and other biological fluids, determination of the impact of environmental factors on levels of a certain body message, assessment of the consequences of a genetic mutation that inactivates a body message or its receptor, and many more.

Observational studies can generate extremely helpful insights but only rarely can determine cause-and-effect relationships. Investigators, therefore, also perform clinical trials, that is, intervention studies in humans. Here again, lots of examples relevant to body message research come to mind, such as assessment of changes induced by administration of a body message, impact of a drug on production of a specific mediator, comparison of two lifestyle interventions on blood levels of a set of cytokines, and many others.

It goes without saying that *in vivo* research—in both animals and humans—is fraught with uncountable ethical issues. While every aspect of scientific research has its own moral quandaries (data falsification, plagiarism, appropriateness of experimental design, conflict of interest, etc.), the additional moral and ethical problems that pertain to *in vivo* studies are so intricate and multilayered that an entire library can be filled with volumes solely dedicated to this issue. The number of weighty, fully legitimate, questions is staggering, ranging from whether any degree of animal suffering should be acceptable in order to benefit human health and the pursuit of knowledge to the safety and vulnerability of volunteers involved in clinical trials, and many more in between. The issue is so complex and controversial that I believe it makes no sense to even attempt to address the ethics of *in vivo* research in a short paragraph other than by emphasizing the lack of easy solutions coupled with the need for strict standards and guidelines.[11]

Sometimes, biomedical research is pictured as a linear progression from *in vitro* studies to *in vivo* investigations, first in animals then in humans. This is indeed the case for clinical testing of new drugs, which can never be administered to humans without prior thorough and appropriate preclinical analysis *in vitro* and *in vivo*. However, nothing of the sort occurs in the research process itself, where the initial observation that sparks a new line of investigation may arise at any level of the chain and proceed in any direction, often mixing *in vitro*, *ex vivo*, and *in vivo* approaches in no specific or predetermined order.

Adding and Subtracting

Intervention studies, either *in vitro*, *ex vivo*, or *in vivo*, can be conceptualized under two types of experimental tactics that I will refer to as addition and subtraction approaches.

When studying the function of a biological molecule using an addition approach, the investigator adds the compound of interest to the investigational system. The researcher may squirt a body message on top of cells cultured in a dish, inject it into an experimental animal, or even administer it to a (consenting and informed) human research subject. What the scientist is doing here is adding the message to the object of inquiry and evaluating the consequences. Typically, the experimental setup includes study of different doses of the molecule of interest (dose-response) and appraisal of the effects over a period of time (time-course). For example, the researcher could inject a group of rabbits with increasing doses of interleukin-1, a cytokine reviewed in the next chapter, and measure the rise of fever over time, the change in temperature brought about when different amounts of the message interleukin-1 are added. In another example, an investigator may inject mice with increasing amounts of the adipokine leptin (discussed in Chapter 7) and then measure how much food the animals consume in the course of the following days, the change in appetite exerted by the message leptin.

The addition method is powerful in providing information about the potential effects of a biological molecule, about what a mediator can do. Researchers obviously need to have at least some clue as to which responses they would want to measure. In the examples mentioned above, the scientist would have to have at least a starting hypothesis that interleukin-1 might cause fever and that leptin could possibly alter appetite, otherwise a totally irrelevant response may be measured and shown not to be affected by the specific message (which can actually be an important finding in its own right).

As powerful as it is in answering the question of what a mediator can do, the addition approach cannot reveal whether the message produced by the animal, human, or cell—the endogenous message—is actually required for a particular change to take place, this method being silent as to what a message is needed for. To answer this other kind of question the researcher needs instead to use subtraction methods. Here, the scientist will block or remove the endogenous message from the system and carefully watch what happens as a result of this subtraction intervention. If we were to use a subtraction approach in the fever experiment described above, we would inject the rabbit with a substance that prevents production or activity of interleukin-1 and observe whether this intervention affects the febrile response to a stimulus (an infection, for instance). In the second example, we would neutralize or remove leptin and observe how much food the mice eat. Thus, subtraction experiments ask whether a specific molecule is necessary for a certain change to take place, in our examples whether endogenous interleukin-1 is necessary for fever to rise in response to infections and whether leptin is necessary for mice to eat a certain amount of food. Here again, a starting hypothesis is necessary for investigators to decide which response they need to measure.

Albeit unfortunately being the cause of much suffering, genetic diseases provide scientists with the opportunity to observe the effect of naturally occurring subtraction approaches. Mutations that inactivate a gene can in fact reveal the function of the corresponding protein. By carefully studying individuals affected by genetic diseases, particularly rare conditions that are characterized by extreme phenotypes (extraordinary obesity, severe inflammation, high fevers, selective metabolic disruptions, etc.), scientists have been able to unveil the function of several biological pathways, including many that involve body messages implicated in regulation of inflammation and metabolism. In the upcoming chapters, we will encounter rare genetic diseases as a source of fundamental discoveries on more than occasion.[12]

To summarize, addition and subtraction approaches provide answers to a complementary set of questions. An addition approach asks, "Which changes can this message effect?"; whereas the question posed by a subtraction study is, "Which changes is this message needed for?" Scientists obviously need to ask and try to answer both types of questions. Whether an investigation begins by adding or subtracting often depends on the available technologies as well the state of the knowledge in the specific area.[13]

Function and Structure

We already mentioned that scientists discovered the vast majority of body messages, including many of those presented in this book, by a function-first approach, that is, by identifying the molecule(s) responsible for a specific biological change. Finding the molecular substrates for control of body temperature (fever) and food intake (appetite) are examples of function-first approaches.

More recently, technical developments—particularly genome sequencing and potent molecular modeling tools—allowed investigators to take a reverse approach, to begin with a gene or a molecule and

then try to understand its function. This procedure has been most successful when researchers identified new genes and proteins in the context of a finite biological setting, by examining the presence of such
gene or protein in a particular tissue, their induction by a specific stimulus or their physical resemblance to structures with an already established function, for example. However, as the upcoming chapters will
clearly indicate, even under these favorable conditions, determining
the function of a newly discovered structure remained much more
challenging than identifying the structure of molecules discovered by
chasing a biological change.

On the other hand, it is also the case that once a new molecule has
been identified within the framework of one specific function (or a few
functions), widening the scope of its participation to apparently unrelated processes may become difficult. As many investigators can tes-
tify, convincing the scientific community that a given body message is
involved in a situation apparently unrelated to that of its original context can be frustrating and time-consuming, often requiring proof that
is usually not demanded from work that fits into existing paradigms.

There are thus advantages and obstacles in following either of the
two paths, in beginning with function or with structure. The various
members of the first family of body messages introduced in the next
chapter will take us on a journey through the evolution of message
discovery that will address this very issue.

3

Evolution of Message Discovery:
The Interleukin-1 Family

INTERLEUKIN-1, IL-1 in short, is among the first (though by no means *the* first) body messages identified as mediators of inflammation. Besides its numerical elegance, beginning our review with IL-1 and its family members provides a historical perspective on the process of body message discovery, on how research in the field changed over the years as technical improvements paralleled, and sometimes even directed, scientific developments.

A Family Snapshot

Biological molecules are often grouped into families based on likeness of their chemical structure. For proteins, similarity of family members can be traced to resemblance of the respective DNA segments, the genes. Each of the eleven members of the IL-1 family shares a key genetic, and therefore amino acid, sequence that bestows a common shape on the proteins.

The eleven members of the IL-1 family are currently called IL-1α, IL-1β, IL-1Ra, IL-18, IL-33, IL-36α, IL-36β, IL-36γ, IL-36Ra, IL-37, and IL-38. The specification "currently called" is important, because names of biological molecules often change together with scientists' understanding of their characteristics. These eleven proteins are the body messages, the ligands, that need to attach to their respective receptors on target cells to deliver information. We know of at least seven receptors that interact with ligands of the IL-1 family, either to activate a response in the target cell or to prevent such activation. In addition to the eleven ligands and their (at least) seven receptors, a plethora of other molecules regulate synthesis and activity of each member of the IL-1 family.[1]

It's easy to get lost in this molecular forest, so let's proceed with some order.

The Founders

"From today's standpoint, the multiple biological activities of IL-1 can be considered the birth of cytokine biology," wrote Dr. Charles Dinarello, a pioneer in the study of IL-1 and currently a professor at the School of Medicine of both the University of Colorado in the United States and Radboud University in the Netherlands.[2]

Multiple biological activities IL-1 indeed does have. In fact, what in 1979 became known as IL-1 entered the world of biomedical research under a multitude of names that reflected the varied scientific interests of the investigators who studied this molecule. Starting out with multiple names has been a common feature in the history of body messages, as we will see over and over again in the upcoming chapters. Another commonality among body messages is the ability to effect disparate—seemingly unrelated—biological changes, a feature that was considered anathema at the time of their initial identification, at the birth of cytokine biology to quote Dr. Dinarello, but is now a bread-and-butter concept.

Three separate function-first avenues ushered researchers to IL-1. Whereas one path involved the search for the mediator responsible for a relatively recondite *in vitro* effect, induction of proliferation of T lymphocytes, the other two roads concerned the quest for the molecular messengers at the basis of *in vivo* processes almost everybody is familiar with: fever and arthritis. Let's begin with fever.

Fever

A famous quote by nineteenth-century Canadian physician William Osler states, "Humanity has but three great enemies: fever, famine and war. Of these, by far the greatest, by far the most terrible, is fever." This may sound strangely apocalyptic in our days of over-the-counter antipyretics, but for millennia fever had been synonymous with deadly infections that decimated humanity, once-mysterious conditions such as plague, typhoid, smallpox, tuberculosis, puerperal fever, and especially malaria, with its characteristic periodic cycles of high temperature.

Why does the body respond to an infection by spiking a fever? Remarkably, after centuries of body temperature recordings and despite many theories about fever's potential function, we don't have a definitive answer to this question. A moderate temperature increase may boost protective mechanisms against microbes, but high fevers actually endanger the organism. As Dr. Tamas Bartfai, a neurobiologist and expert in fever research told me, "While we learned much about how fever is brought about—what the cellular and molecular mechanisms are—we still don't have an answer about what fever is 'good for,' if it is good for anything." Even though we have yet to grasp fully the evolutionary meaning of fever, we have reached a rather detailed understanding of the mechanisms that led to its induction, as Dr. Bartfai stated.[3]

It was in trying to answer the apparently simple question "What causes fever?" that researchers who chose to walk the fever path discovered the founders of the IL-1 family.

Not every fever has a microbial origin but infectious agents such as bacteria and viruses are most commonly involved. It would thus appear obvious to conclude that the microbes themselves cause fever. Indeed, experimental animals, particularly rabbits, do mount a fever when exposed to microbes, just like infected humans. Researchers can even induce fever by injecting rabbits with microbial extracts, obtained by killing bacteria and extracting their components, showing that the bugs do not need to be alive, or even whole, for fever to develop. However, it takes time—usually a few hours—for temperature to rise after injection of microbes or their products into a rabbit (or a human), suggesting that something needs to happen inside the body before the temperature spike occasioned by those microbes can take place. Could it be that cells in the body sense the presence of microbes, no matter of which type, no matter if dead or alive, and in response produce a substance that conveys the message "Start a fever"?

Investigators in the 1940s and 1950s began testing this very hypothesis using various approaches. In one of many such studies, Drs. Elisha Atkins and W. Barry Wood injected a set of rabbits with dead bacteria, which the investigators already knew would cause fever. At fixed intervals after inoculation, the researchers drew blood from the rabbits, made sure the dead bacteria were no longer present, separated the serum (the liquid part of blood), and injected it into a new set of naïve rabbits that had not received the bacteria, the recipients. Atkins and Wood were working under the hypothesis that cells in the body of rabbits inoculated with the dead bacteria would synthesize a substance able to induce fever in recipient animals, even though these latter had never encountered those bacteria. Indeed, recipient rabbits rapidly spiked a fever when injected with serum from animals that had been inoculated with bacteria, and they did so quickly, without any delay. This indicated the serum they had received contained a substance, produced by cells of the rabbits inoculated with bacteria, which carried across the message "Fever!" to rabbits that had never seen those microbes. This body message was designated as an endogenous pyrogen, to signify that it was made by the body's own cells (endogenous) and that it induced fever (pyresis, hence pyrogen).

Additional studies demonstrated that leukocytes—the white cells of the blood—were the main producers of this endogenous pyrogen, which was therefore also assigned the alternate name of leukocytic pyrogen. By the mid-1970s, researchers had learned that leukocytic pyrogen did not exist as a preformed substance inside cells, but that leukocytes instead synthesized it after exposure to microbes. This finding explained the delay in fever development when investigators injected rabbits with bacteria: it took a few hours for leukocytes to sense the presence of microbes and set in motion the cellular machinery leading to generation of the pyrogenic message. Scientists soon determined many of leukocytic pyrogen's biochemical characteristics, but the exact identity of the fever-inducing message for the moment remained unknown.[4]

Arthritis

The other *in vivo* process that led researchers to IL-1 is the chronic inflammatory disease of the joints known as rheumatoid arthritis, a disorder of unclear origin that mostly affects women in young adulthood–middle age and is not to be confused with the most common form of arthritis (osteoarthritis), which is instead a degenerative condition caused by wear and tear. As an 1880 quote by British surgeon William Ord indicates, rheumatoid arthritis has been puzzling the medical world for quite a while, "A definition of rheumatoid arthritis being required, it may be stated as follows: a persistent or progressive inflammation of one or more joints, in origin neither rheumatic, nor gouty, nor scrofulous, nor in any way specific."[5]

In individuals with rheumatoid arthritis, the lining of the joints, the synovium, grows to form a thick layer of highly inflamed tissue that progressively destroys cartilage and bone, causing such pain and deformity that movement becomes all but impossible. The small joints of hands, feet, and the top of the spine are mostly affected, but other joint sites as well as various organs (such as skin, kidneys, and lungs) can be damaged, too. Although we do have some hints, unfortunately science has not made a huge progress from Dr. Ord's writing as far as understanding the ultimate causes of rheumatoid arthritis. However, lack of understanding of rheumatoid arthritis' origins does not mean we are helpless in confronting this dreadful disease. Development of pharmacological treatments for this chronic inflammatory condition is one of the most thunderous achievements of cytokine research, and we will encounter this disease in several chapters, since more than one body message is involved in its pathogenesis.

The second route that led to the discovery of IL-1 took the form of understanding the mechanisms responsible for the devastation in the joints of rheumatoid arthritis patients. In parallel with the fever studies mentioned above, and using conceptually similar approaches, scientists in the mid-1970s demonstrated that leukocytes present in the inflamed synovium of diseased joints produce and release substances that set in motion processes that eventually destroy bones and carti-lage. Among the first such demonstrations was the work of Dr. Stephen Krane and collaborators at Harvard Medical School, who in 1977 pointed to proteins made by leukocytes as the factors responsible for joint inflammation and damage. The biochemical characteristics of one of these factors were similar to those of the leukocytic pyrogen that fever researchers were studying at the same time, but here too the mediator's identity remained obscure for the time being.[6]

Finding IL-1

Not being in possession of IL-1's official identification card, the exact nature of the mediator under investigation still being unknown, scien-tific detectives of the mid-1970s relied on clues obtained in the pre-vious decade to single out this body message among the dense molec-ular crowd present in the blood of rabbits injected with microbes, the joints of arthritic patients, or the supernatant of leukocytes cultured *in vitro*. Investigators knew IL-1's approximate size and understood some of its other chemical properties. They could also detect and quantify IL-1's presence by assessing the effect of their preparations on lympho-cyte proliferation *in vitro*, damage to cartilage *ex vivo*, or fever in rab-bits *in vivo*.[7]

How did investigators measure fever in a rabbit in the 1960s and 1970s? Here is Dr. Dinarello, who led the team that identified IL-1 as the endogenous (or leukocytic) pyrogen, beginning his studies when he was still a medical student, "First thing I did was to learn how to put a rectal thermometer in a rabbit's ass and let it stay there. And that was part of my life for thirty years. I think I spent equal time in front of the rabbits injecting them and behind the rabbits making sure that they didn't push out their probe."[8]

What on earth could motivate a bright medical student to spend time learning how to stick thermometers into rabbits' behinds?

There is nothing special about my life compared to other scientists I know, except I was very drawn to music. I loved singing; my mother and an aunt would take me to the opera in Boston when I was young. I had to take piano lessons but I also sang in the church choir. I remember having to sing at funerals in the Roman Catholic Church. The body was rolled in on this big black casket and the choir would sing the Dies Irae.

Then I went to high school and I sang in the glee club. One day I had to sing a solo in one of the performances and my father and mother were there. Afterwards, my father said to me, "Look, you are going to starve to death if you think you are going to make it in music. I think you better continue with your interest in chemistry."

So I went to Boston University, where I was a chemistry major, which served me very well later on. I really wanted to be a chemist, I even won the chemistry prize, but under the influence of my classmates I decided to go to medical school.

I went to Yale University, the only medical school in the United States that required a thesis. At Yale, there were two very famous people who led research on fever, Elisha Atkins and Paul Beeson. These were big names in the 60s. I arrived at Yale in 1965, and you had to pick a thesis project. I had a lecture by Elisha Atkins on fever and I got very interested. I asked if I could do my thesis with him and that's how I got into the fever project. So, what did I do? First thing I did was to learn how to put a rectal thermometer in a rabbit's ass."

Dr. Dinarello's thesis, which won him a $2,000 prize at Yale that he used "to buy a couch, a desk, and a table for a studio apartment in Boston before beginning an internship in pediatrics at Massachusetts General Hospital," was on the production of endogenous pyrogen by liver macrophages, also known as Kupffer cells. As mentioned in Chapter 1, macrophages are immune cells that are particularly good at eating bacteria and the debris of life. These cells are also the strongest producers of some of the most powerful cytokines and other body messages. Macrophages keep our bodies healthy by clearing infections and maintaining the integrity of our tissues, but can become damaging when improperly activated, as in rheumatoid arthritis and other inflammatory conditions.

During his medical internship and residency, Dr. Dinarello was itching to get back to the laboratory. "The Vietnam War was on and I had to either go to Vietnam for two years or trying to get into the Public Health Service, in which you had to do three years. I opted for that one, obviously. I was interviewed by Sheldon Wolff, who ran the other big lab at the time working on endogenous pyrogen, and he took me in based upon my thesis, which had been published. So I arrived at the NIH in 1971, and sat in Shelly's [Dr. Wolff] office. He said, 'Are you going to continue working on Kupffer cells?' I said, 'No, I want to purify this molecule.' Except I had no idea how to purify a molecule."[9]

Purifying a molecule means fishing it out among the many others that float around either in blood or, in this case, in the supernatant of cultures of human monocytes stimulated with bacterial products *in vitro* (recall that macrophages are named monocytes when they are present in blood), the experimental system Drs. Dinarello and Wolff had chosen to purify the pyrogenic message. Several techniques were available to accomplish this task in the 1970s, but they all had a problem: the protein you were trying to single out had to be present at rather high levels, which was definitely not the case here. Because of its extreme potency, even the most productive cells (the monocytes Dr. Dinarello was using) synthesize and release very little of what was then still called leukocytic pyrogen, so this molecule's concentration was extremely low even under the best possible experimental conditions (this could be fully appreciated only afterward, at the time it was simply frustratingly incomprehensible). The low levels of leukocytic pyrogen made purification a daunting task, because the techniques available were powerless. In fact, many researchers were actively attempting to isolate and purify the same molecule during those years, and all were having a really hard time.

Rabbits came to Charles Dinarello's rescue. After several unsuccessful attempts using most of the traditional biochemical techniques, together with Dr. Wolff he decided to try to go about it in a different way. They concentrated their crude preparation of human leukocytic pyrogen and repeatedly injected it under the skin of rabbits, to push the animals to react against the foreign substances present in the mixture—which came from human cells and therefore were not familiar to rabbits—with production of antibodies. This is the very same principle of vaccination: the immune system reacts against an unfamiliar substance by producing antibodies that specifically recognize, and potentially neutralize, the invader. Rabbits' antibodies were crucial in allowing Dr. Dinarello and his collaborators to come up with ever-cleaner preparations of leukocytic pyrogen, which were then patiently tested for their ability to induce fever in rabbits.

These efforts eventually panned out. In 1977, Drs. Dinarello and Wolff along with Dr. Lois Renfer reported isolation of pure human leukocytic pyrogen. In 1979—when the term *interleukin* itself was coined—the protein received its current designation, interleukin-1.[10]

Soon afterward, the fever and arthritis routes of discovery converged. In 1980, Drs. Jeremy Saklatvala and John T. Dingle, in Cambridge, England, reported isolation of a protein from *ex vivo* cultures of joint synovial tissue extracted from freshly slaughtered young pigs, a protein with the ability to degrade cartilage obtained from bovine nasal septum. Under these conditions, degradation of cartilage is caused by the breakdown of large biological molecules into their individual components, a process that is classified under the general term of *catabolism*, the arm of metabolism that breaks things apart.[11] Saklatvala and Dingle had this process in mind when they named the protein they had discovered catabolin based on its tissue-destroying abilities, a clear instance of the worlds of inflammation and metabolism coming together.[12]

Dr. Saklatvala soon demonstrated that catabolin was the very same protein as IL-1, with a similar fate awaiting the leukocyte mediator discovered a few years earlier by Dr. Krane's group at Harvard while also studying arthritis.

By the early 1980s, thus, researchers had identified the protein message behind fever and arthritis. However, when dealing with a protein, obtaining its ultimate ID requires deciphering its genetic sequence among the thousands of genes in our DNA. This was even more important for IL-1, because there were claims by several researchers that molecules other than IL-1—or even microbial contaminants, particularly one called endotoxin—were the actual cause of fever, an issue that was hotly debated and that caused much grief to Dr. Dinarello. "Endotoxin was my big struggle," he told me. "During this period, people would say endogenous pyrogen and IL-1 cannot be the same molecule; they were so convinced that the fever molecule was endotoxin bound to an endogenous protein. Somebody even wrote a whole paper about this. It was actually that article that forced me to go into this issue. We had to do the cloning to prove us right."

The term *cloning* has multiple definitions. In this context it means decoding the genetic sequence that corresponds to the amino acids of the IL-1 protein, artificially reproducing that sequence in the laboratory and then turning it into the IL-1 protein itself, eventually demonstrating that this manufactured IL-1 has the same activity—carries the same message—as the natural protein. Now that the human genome (and those of many other living and extinct beings) has been decoded, finding that sequence and manufacturing that protein would be a breeze for those with the necessary expertise. But it wasn't until 1978 that a gene, the one for the hormone insulin, was ever cloned. Embarking on cloning IL-1 in the early 1980s can at best be described as intimidating.

There Is no Beta without an Alpha

"Together with Alex Rich and postdoctoral fellow Phil Auron, we started the cloning project at the Massachusetts Institute of Technology on February 1, 1982, and I entered the first experiment into my notebook." This is again Dr. Dinarello, speaking about the initial steps in identification of the IL-1 gene. Up until now, together with his intense scientific interests he had somehow managed to keep up his musical career, becoming part of the chorus that sang with the Boston Symphony Orchestra. He now had to make the fateful decision of choosing between music and science. "How I mixed singing with science I still cannot explain. But I did it; it was like my second life," he told me. "But one day, during the cloning project, I asked Phil Auron if he would destain a Western blot for me.[13] I had to be at the rehearsal for performance of Tosca with the Boston Symphony Chorus, and the rules were if you skipped two rehearsals in a row you had to reaudition, and auditions were very difficult. Phil screamed at me, 'You have to make a decision if we are going to do this project or you are going to sing.' At that point I realized I really could not do both, the cloning project had become very intense, everything had to be done by scratch, so I did not go and I missed two rehearsals and did not reaudition."

After almost two years of intense efforts working with human monocytes obtained from leftovers at the local blood bank, the investigators eventually managed to decipher the genetic sequence of IL-1, deduce its amino acid composition, prepare a manufactured IL-1, and demonstrate its biological activity. The group's accomplishments clearly indicated that an endogenous protein, the body message IL-1, could induce fever, thus cooling down debate about the role of microbial contaminants and additional proteins. A few months later, the form of IL-1 cloned by Dinarello's group received the designation of IL-1β.[14]

For something to be called a β, an α must first exist. Already in 1974, during their early purification efforts, Drs. Dinarello and Wolff had demonstrated that leukocytic pyrogen comes in two forms, with different molecular sizes and chemical characteristics. They wisely decided it would be enough of an effort to chase one of them, the one that took the name IL-1β. Other investigators went after the other form, following for the most part the road of the lymphocyte-stimulating factor (with occasional dips into the field of arthritis). In collaboration with scientists at the pharmaceutical company Hoffmann-La Roche, Dr. Steven Mizel led the team who first purified and then cloned the other kind of IL-1—which was later called IL-1α—starting with a mouse macrophage-like cell line.[15]

Studies reporting the cloning of both forms of IL-1 were published at the tail end of 1984: the Mizel article on mouse IL-1 in the journal *Nature* at the end of November, the Dinarello report on human IL-1 in the *Proceedings of the National Academy of Sciences USA* at the beginning of December. About six months later, in June 1985, a third study in *Nature* announced the cloning of both forms of human IL-1, introducing for the first time the terms IL-1α and IL1β. The authors of this third report were employees of Immunex, a biotech company in Seattle, who had just submitted their sequences to the United States Patent Office to obtain exclusive rights on the discovery.[16]

Though IL-1β was discovered by (mostly) chasing fever and IL-1α (mostly) via the lymphocyte proliferation route, early on it was already evident that the two cytokines exerted overlapping, almost indistinguishable effects: indeed, both could cause fever when injected into rabbits, degrade joint cartilage and make lymphocytes proliferate *in vitro*.

A few years after cloning of IL-1α and IL-1β, researchers at Hoffmann-La Roche and, separately, at Immunex identified the IL-1 receptor, the protein IL-1 needs to attach to in order to transmit a signal inside the target cell.[17] These studies clearly demonstrated that IL-1α and IL-1β use the same receptor to deliver their information, so it was no wonder their messages coincided. Additional studies eventually showed that binding of either IL-1α or IL-1β to the receptor on the surface of a receiver cell elicits a series of intracellular events, later sorted out in great detail, that change that cell's behavior. Importantly, the nature of the change effected does not depend on which form of IL-1 is carrying the message but on the type of cell that to that message responds: a T lymphocyte proliferates, enhancing immunity; a cell in the brain releases prostaglandins, causing fever; a macrophage in the joints produces enzymes that degrade bone and cartilage, exacerbating arthritis. Many other cells in the body react to IL-1 in countless additional ways, with an overall inflammatory response that helps the body fight infections but can also, by itself, cause disease.

However, just because the activities of the two forms of IL-1 can overlap, it doesn't mean they do so in the real world. What a message can do is a separate issue from what a message is needed for, as discussed in Chapter 2. Even though they are both body messages, IL-1α and IL1β behave in noticeably different manners. Whereas IL-1α is present in most cells at all times, IL-1β is instead almost exclusively produced by leukocytes and is made only when these immune cells are responding to danger, which can take the form of an infection, autoimmunity, or various other circumstances. While IL-1α mostly stays inside the cells that make it, even reaching into their very core, the nucleus, cells that synthesize IL-1β have a complex machinery that, when activated, throws this cytokine out so it can deliver its message to other cells.

As we have seen, identification of the two forms of IL-1 followed function-to-structure approaches, the most common way of finding body messages. Many scientists, from different fields of research and various work settings, participated in the quest. Despite the occasional disagreement, collaborations and free exchange of ideas, together with sheer determination, made all of this possible. But science is not always a bed of roses, especially when money gets in the way.

The Dark Side of a Bright Discovery

"Sir, at the Fourth International Lymphokine Workshop in West Germany on 17–21 October 1984, an executive of the company Immunex challenged our claim to have cloned the complementary DNA for interleukin-1 (IL-1) and informed the workshop that we had instead cloned some other monocyte product." Thus begins a letter to the editor of *Nature* published in 1986 and signed by Drs. Wolff, Auron, Dinarello, and Lanny Rosenwasser, authors of the article that—a couple of months after that 1984 workshop—had first reported cloning of human IL1β.[18]

During the workshop, which took place at the historic palace Schloss Elmau, one of Immunex's two cofounders had startled the audience by publicly questioning the validity of the IL-1 sequence presented by Dr. Auron, without caring to offer any data in support of his claim. However, in June 1985, just a few short months after having publicly challenged the IL-1 sequence found by Dinarello's group, Immunex scientists had published a paper in *Nature* reporting the sequence of both human IL-1α and IL-1β. Since the IL-1β DNA sequence included in this *Nature* article overlapped with the one the Immunex executive had questioned at Schloss Elmau, Drs. Wolff, Auron, Dinarello, and Rosenwasser were justifiably surprised. They decided to voice their astonishment publicly in the letter to *Nature* quoted above: you first make a big fuss telling us we have the wrong genetic sequence and then publish that very same sequence! What kind of behavior is this?[19]

The accident at the time appeared to be nothing more than an annoying strife between two competing teams. "We just assumed that they had eventually cloned it themselves," Dr. Dinarello told me during our interview. Not so, as it turned out some ten years after the events described above.

When researchers are ready to report the results of their studies, they write them up in the form of a report, which they submit for publication to one of many scientific journals. Some of these journals carry greater cachet than others, being accepted inside their covers constituting recognition of the importance of a scientist's research findings. The journal *Nature* is way up there, the very height of prestige with a few equals. When a research report is submitted to a journal for publication, the journal's editors send it out for peer-review. This means that two or three investigators not connected with the study, and supposedly not in competition with the study's authors, review the report to check its accuracy, novelty, and potential impact. Reviewers then submit their opinion and recommendation to the journal's editors, who decide whether to accept the report for publication. If the article is rejected, the authors usually submit it to another journal and the cycle starts all over again. Obviously, the whole process is supposed to be highly confidential.

"We had first submitted our paper to *Nature* in May of 1984; the manuscript was rejected in July," Dr. Dinarello told me referring to his IL-1β-cloning article. October is when the public accusation of inaccurate results at the German workshop by the Immunex executive occurred, with the Dinarello team meanwhile having resubmitted their report to the *Proceedings of the National Academy of Sciences USA*, the journal that ended up publishing it in December 1984. In June 1985, *Nature* published the Immunex paper, which contained the very same IL-1β sequence previously reported in the Dinarello article and questioned by the Immunex executive at the Schloss Elmau workshop.

Years went by. The Dinarello group and Cistron, the small biotech company they were working with to develop the discovery of IL-1β into pharmacological applications, began a lawsuit about the ownership of the patent for the right to commercialize IL-1β, which had been assigned to Immunex. "A lawyer told us the IL-1β sequence in the Immunex article was not the same sequence in the Immunex patent," Dr. Dinarello explained. The IL-1β genetic sequence is a string of approximately fifteen hundred ACTG, the letters of the DNA alphabet. Among these many letters, the original sequence reported in the Dinarello group's 1984 article contained nine mistakes, the equivalent of nine typos in about a single-spaced page completely filled with the same four letters. The lawyer had patiently compared—letter by letter—the three IL-1β sequences: the first in time being the one present in the Dinarello article (which had initially been submitted and rejected by *Nature*), the second in time represented by the sequence present in the patent submitted by Immunex, and the third represented by the sequence published in the Immunex 1985 *Nature* paper. The very same nine errors present in the Dinarello sequence were also present in the sequence of the Immunex patent, but these mistakes were then corrected when Immunex investigators published their own article in June 1985. "The lawyer found the difference; we just assumed that the Immunex patent was the same as their paper. Am I going to look at fifteen hundred nucleotides? Some lawyer did it! It's amazing that he did it," Dr. Dinarello exclaimed.

The same nine mistakes in two genetic sequences that were supposedly independently generated represented way too much of a coincidence, so a legal investigation was initiated, producing startling results. When Dinarello's group had initially submitted their article to *Nature*, an Immunex researcher was among those chosen for peer-review. That scientist's report to *Nature*'s editors strongly contributed to rejection of the Dinarello manuscript, but the IL-1β genetic sequence the manuscript contained was copied—together with the nine mistakes—and used for the Immunex patent application. "Immunex had a $16 million contract with the pharmaceutical company Syntex if they were the first to clone IL-1," Dr. Dinarello explained. "So they rejected our paper, told Syntex they had cloned IL-1—here is the patent—and Syntex gave Immunex $16 million. But they didn't have any IL-1 sequence, they only had our sequence. Only afterward did they generate their own IL-1β sequence and correct those nine mistakes."

The lawsuit was eventually settled and the Immunex patent for IL-1β turned over to Cistron. "During the trial, Immunex employees testified that they had had a meeting saying, 'When we go to Schloss Elmau we would stand up and say it's not IL-1.' They actually had it planned! They planned to do that!" an emotional Charles Dinarello told me in 2014, these events still vivid in his memory after four full decades.[20]

As useful as it is, peer-review, like other similar endeavors, can only function on the honor system, when scientists declare their conflicting interests and refrain from taking undue advantage of confidential information. Adding money to the mix certainly does not help. The story of IL-1's cloning is unfortunately not the only example in which the ugly side of competition showed up in the world of body messages. The success of this field of investigation in translating the results of basic science into therapies that are helping thousands of individuals, but also generating lucrative profits, has every so often unmasked behaviors that are utterly obnoxious when not outrightly illegal.

The Antagonist

As we discussed a few pages back, scientists identified IL-1 by searching for the message responsible for specific biological changes: fever, arthritis, lymphocyte proliferation. It was the absence of these changes, even when IL-1 was supposedly present, that then drove investigators to the next member of the IL-1 family, IL-1 Receptor antagonist, or IL-1Ra. "Also the absence of reaction is a reaction," Italian physician-philosopher Giorgio Prodi once stated.[21] If you know something should happen in your system but it doesn't, you better take notice. At first you might think it's a fluke, some technical error that you try to fix. But when you observe the same phenomenon over and over again, it means you are on to something.

Would you be comforted or annoyed if you knew that a researcher on another continent was pondering your same questions, facing the very same absence of an expected reaction? Working near the gorgeous peaks of the Swiss Alps and the Rocky Mountains, two rheumatologists with research in their heart—Drs. Jean-Michel Dayer and William Arend—independently found IL1Ra by observing absence of expected IL-1 activity.

Now formally retired but still active, Jean-Michel Dayer was born in Lausanne, Switzerland, and grew up in Sion, in the alpine Swiss canton of Valais. After studying Latin and Greek in high school, he continued with medicine, finding himself deeply interested in inflammation and wound healing, perhaps as a result of having suffered from chronic arthritis of one knee since he was a little child. "Arthritis and other connective tissue diseases were very mysterious at the time, without clear mechanisms," he explained during our interview. At the end of his medical training, in the mid- to late 1970s, Dayer spent a few years at the Arthritis Unit of Massachusetts General Hospital, part of Harvard University. It was during this period that, working with Dr. Stephen Krane, he was among the first to recognize the existence of the joint-destroying factor that later turned out to be IL-1, as we just reviewed.

In 1981, Dr. Dayer returned to Switzerland, at the Faculty of Medicine in Geneva. "Since laboratory facilities in Geneva were very limited, I got some space in a small temporary wooden building near the hospital and had to transform the toilet into a cell culture room to have at least a closed space to avoid contamination! My position was at that time research assistant professor, with some clinical duties. Even though I had been assistant professor of medicine at Harvard Medical School, when I returned to Switzerland I had to wait, show my abilities, and start again the whole promotion path. I was eventually promoted all the way to full professor and continued working on the biology of inflammation and arthritis."[22]

Like Jean-Michel Dayer, William Arend also spent most of his childhood near mountains, on a farm at the foothills of the Adirondacks. "My early education was carried out in the schools of my village, where I graduated as valedictorian. My activities outside of school included hiking, fishing, canoeing, skiing, and generally spending time in the outdoors. My family had a camp on a small lake in the Adirondacks where there was no road, electricity, or running water. We carried in our supplies in backpacks and enjoyed vacation in summer and winter," Dr. Arend reminisced. He always enjoyed science and mathematics; his grandfather was a physician and young Bill liked spending time with him in his office. After majoring in chemistry at Williams College in Williamstown, Massachusetts, Arend attended medical school at Columbia University in New York City. "Because of the high quality of the medical training and the ready access to skiing and other activities in the outdoors I moved to Seattle in 1964 for internship and residency in internal medicine at the University of Washington. My clinical medical training was interrupted by the Vietnam War and I spent two years of military service in Washington, DC," Dr. Arend continued. He then entered a clinical and research fellowship in rheumatology, stimulated by educators who had raised his interest in the field. In 1970, he started an independent laboratory and remained on the faculty of the University of Washington School of Medicine for ten years. In 1980, with a fellowship from the Guggenheim Foundation, Arend spent a year doing basic science at the Strangeways Research

Laboratory in Cambridge, England, where Drs. Jeremy Saklatvala and John Dingle had just discovered catabolin, one of the many arthritis-related incarnations of IL-1. When he returned to the United States, Arend continued working on IL-1, for two years at the University of Texas at Houston, then at the University of Colorado School of Medicine in Denver, where he remained for the next thirty years, some of them as head of the Division of Rheumatology. Now retired, he continues to enjoy his boyhood love of the mountains by hiking, backpacking, and skiing (downhill, cross-country, and backcountry).[23]

Two not dissimilar stories, with both physician-scientists crossing the Atlantic, in opposite directions, to gain experience in studying the body messages involved in rheumatoid arthritis. Moreover, both scientists had been involved with early research on IL-1.

To better appreciate the process by which absence of IL-1 activity occasioned identification of IL-1Ra by Drs. Dayer and Arend, a brief excursus into the field of biological measurements can be of help.

Quantity versus Activity

When police stop a driver who is behaving erratically, they usually do a couple of things to check whether alcohol is involved. They use a breathalyzer, a device that measures the level of alcohol in the driver's breath, and they often also observe whether the driver is able to walk a straight line. The first test measures quantity, that is, how many molecules of alcohol are present in someone's breath. The second test measures activity, that is, the effect alcohol has had on the driver's ability to walk straight. Many variables can influence this ability, including what else is present in the driver's system (was there eating or taking other drugs together with drinking?) and how sensitive the driver is to alcohol (is this a habitual drinker or a newbie?). Where the breathalyzer is specific, being concerned only with measuring alcohol and completely ignoring other substances, the walking test is far less specific, since many factors can affect someone's ability to walk straight. But a less specific test can at times be more relevant than a specific one if the goal is road safety. In this example, the walking-straight test is the equivalent of a bioassay, a measurement of a molecule's ability to effect a biological change, while the breathalyzer corresponds to a test that quantifies the amount of a given molecule, irrespective of whether this molecule is active, of whether it is able to effect a change. Which of the two tests is more appropriate depends on the specifics of the question the investigator is asking, since bioassays and quantification assays provide answers to complementary questions.[24]

In 1977, US physicist Rosalyn Yalow received the Nobel Prize in Physiology or Medicine for development of the radioimmunoassay, a quantification method that has been essential in advancing biomedical research and that, together with its descendants, is still widely used today. These simple and rapid techniques allow for precise quantification of the amount of any given molecule in blood, urine, supernatant of cells cultured *in vitro*, or other kinds of samples. The results of these assays tell us how much of a specific, individual molecule (say, IL-1β) is present in that sample, but do not provide any information on whether the molecule being measured is able to effect a change in its target, whether the molecule indeed acts as a message. Even when many molecules of a mediator are present, their ability to effectively deliver their message may be hampered for a variety of reasons. Other mediators may simultaneously deliver an opposite message, thus nullifying the end result, cells may not have enough receptors on their surface to sense the message, or—even if receptors are present—inhibitors can prevent the message from activating target cells.

To measure the activity of a body message like IL-1 we need a bioassay, a test that quantifies the actual change the message brings about. This change can be fever in a rabbit injected with a sample that contains IL-1, damage to cartilage exposed to fluid containing IL-1, or proliferation of lymphocytes cultured *in vitro* in the presence of IL-1. The magnitude of the change can be used to gauge how much IL-1 is present in the starting material: more fever, more cartilage damage, more cell proliferation equals more IL-1 activity (though, for the reasons just discussed, this does not necessarily equate to more IL-1 molecules).

Implementation of bioassays is typically complex and finicky, since these tests imply use of living matter to assess the desired biological effect. Bioassays take considerable time, effort, and skill. They are also somewhat imprecise, since several known and unknown substances present in the sample can interfere with the outcome. Although this lack of precision can be a disadvantage, it may also turn around to be an asset, as we are going to see for the discovery of IL-1Ra.

Bioassays are definitely not as neat and easily packaged into ready-to-use kits as quantification tests, and in large part they have fallen out of favor as means of measuring body messages. But the information bioassays provide is invaluable, and complementary to that obtained using a quantification test. If the message is the change, as we discussed in Chapter 1, then evaluating the activity of a body message—its ability to effect that change—should be an essential part of its measurement. Alas, this is rapidly becoming old knowledge that is no longer considered cutting edge, but should still remain an important part of a scientist's armamentarium.[25]

Finding Absence

Although in 1977 Dr. Dinarello's team had developed a specific IL-1 quantification test using the technique initially devised by Rosalyn Yalow, for several years the gold standard for measuring IL-1 remained an *in vitro* bioassay, the measurement of proliferation of lymphocytes obtained from the thymus of a mouse (thymocytes). In 1981, that bioassay allowed Drs. Dinarello, Rosenwasser, and Wolff to demonstrate that serum obtained from humans in whom the investigators had induced fever by administration of the bacterial product endotoxin contained not only IL-1 but also a factor that suppressed IL-1-induced thymocyte proliferation *in vitro*. Although other investigators independently noted the presence of an IL-1 suppressor activity under various circumstances, nobody managed to come up with this putative inhibitor's molecular identity and mode of action.[26]

It was bioassays that led Drs. Arend and Dayer to the IL-1 inhibitor IL-1Ra, one arriving by way of *in vitro* studies with cultured monocytes, the other through analysis of patients' samples.

During his year in Cambridge, Dr. Arend had learned how to obtain and culture cartilage cells to measure IL-1 bioactivity. Upon returning to the United States, he began to investigate factors that might regulate production of IL-1. To this aim, researchers in his team cultured leukocytes (which contain monocytes, the strongest producers of IL-1) under multiple *in vitro* experimental conditions. They then used bioassays (measuring both proliferation of mouse thymocytes and alterations of rabbit cartilage cells *in vitro*) to test how much IL-1 activity was present in the medium that bathed those leukocytes, supposedly a measure of how much IL-1 the leukocytes themselves had synthesized and released. Surprisingly, certain experimental conditions that should have induced production of IL-1 did not show any IL-1 activity in the bioassays, eventually leading Dr. Arend to speculate that an inhibitor of IL-1's activity might be present under these specific *in vitro* conditions. Arend and his colleagues published their initial findings in 1985.[27]

Meanwhile, Dr. Dayer in Switzerland found absence of IL-1 activity following a completely different route, one that started with analysis of pathological samples. "The discovery of IL-1Ra cannot be dissociated from efforts to purify IL-1 and was the consequence of a surprising observation. In search of having large amounts of biological material to purify IL-1, I thought that patients having lots of monocytes, such as those with monocytic leukemia, would be an ideal source of material. The surprise was that no IL-1 biological activity could be detected in urine and serum of such patients. The crucial experiment was the idea that the biological activity must be present, but masked," Dayer told me, explaining the thought processes that led his team to discover the family antagonist IL-1Ra.[28]

Both investigators landed on a similar approach to try to figure out what was going on in their respective experimental systems. They would take samples where they thought an IL-1 inhibitor might be present, mix them with preparations containing known amounts of IL-1, and use bioassays to test whether the samples would inhibit the activity of the IL-1 they had added. After performing real tour-de-force studies that included hundreds of complex bioassays, the two rheumatologists came to the same conclusion: their samples indeed contained a substance—a protein—able to inhibit the activity of IL-1 in a very specific manner, without altering other cytokines' effects.

The activity of a body message such as IL-1 can be inhibited in multiple ways; it was therefore necessary to understand the mechanism underlying the observed inhibitory function. In collaboration with other investigators, Dr. Dayer demonstrated that the inhibitor, concentrated from the urine of febrile patients, attached to the very same receptor used by IL-1 but did not activate the receptor like IL-1 did. Instead, the new protein occupied the place of IL-1, preventing this cytokine from binding to its own receptor: the new protein thus acted as a classical receptor antagonist. This was the first example of a natural, endogenous antagonist that worked by blocking the binding of a cytokine to its receptor, an accomplishment made possible by the dedication of a member of Dr. Dayer's team, Philippe Seckinger, a young, brilliant, and talented PhD student who sadly passed away in 1996. Dr. Arend's team later independently confirmed the receptor antagonist properties of the IL-1 inhibitor.[29]

Working at the University of Colorado in Denver, in 1987 Dr. Arend entered into a scientific collaboration with Synergen, a biotech in nearby Boulder, with the goal of purifying, sequencing, cloning, and manufacturing the recently discovered IL-1 inhibitor. His group obtained monocytes from blood of (consenting) medical students and cultured these cells to generate liters of supernatant, which were then transported to Boulder for isolation of the inhibitor. After two years of effort, Dr. Charles Hannum and his colleagues at Synergen obtained enough purified material to partially sequence the protein, revealing the presence of a new cytokine with some sequence similarity to IL-1α and IL-1β. Their 1990 report confirmed that the inhibitor worked as a specific receptor antagonist of IL-1 and introduced the name the cytokine still holds, IL-1 Receptor antagonist, or IL-1Ra. Cloning soon followed—in large part thanks to the skilled work of Dr. Stephen P. Eisenberg—together with a patent.[30]

The work leading to the discovery of IL-1Ra followed almost parallel paths, as we have seen. How did the two rheumatologists feel about the competition? Dr. Arend told me how he saw Jean-Michel Dayer at scientific meetings between 1985 and 1990, but that they never discussed their respective interests on the IL-1 antagonist. "I knew about our competitors, and there may have been other groups as well not known to me at the time. Scientific competition is always present and should be thought of as stimulatory, not annoying," he said. Dayer, while stressing that his group was the first to demonstrate the receptor antagonist mechanism of IL1Ra, mentioned that, retrospectively, their team in Switzerland was more interested in science and academic research than in business, unlike in the United States, and probably naïvely talked too openly at international scientific meetings or even with people working in industry. "We also could not compete with all the excellent scientists at Synergen who did the final work and were then taken by Amgen [in 1994, the biotech Amgen acquired Synergen]. We were not in the position to have a patent since the publication on ligand-binding inhibition was published. This was unfortunate," Dr. Dayer concluded. A look at the current clinical use of ILıRa can easily explain why he feels this way.[31]

Clinical Developments

After cloning IL-ıRa, scientists at Synergen (and then Amgen) started to produce the cytokine in recombinant form. It's called recombinant a protein that is artificially made by introducing a piece of DNA that corresponds to that protein's genetic code inside organisms, generally bacteria, which are themselves called recombinant when so manipulated. Recombinant bacteria grown inside huge vats, similar to those used in breweries to make beer, will transcribe the DNA sequence into RNA and eventually translate it into the protein of interest, which is then isolated, concentrated and purified. This technology is what allows manufacturing of insulin for treatment of diabetes, of vaccines against Hepatitis B, and of all the protein-based modern therapeutics used to treat diseases ranging from cancer to autoimmunity, including IL-ıRa.

Because IL-ıRa antagonizes the activity of IL-ı, scientists could now use recombinant ILıRa as a subtraction tool to understand the role of IL-ı in many contexts, *in vitro*, *ex vivo*, and *in vivo*. Clinical trials in humans were also initiated. Sepsis, a life-threatening complication of infections that has an extremely high mortality rate and no available cure, was the first condition in which IL-ıRa was tested in a clinical setting. Unfortunately, after initial promising findings, the highly disappointing results of a large clinical trial were published in 1997, showing that IL-ıRa failed to improve survival in septic patients. However, trials continued in other disorders and positive outcomes soon started showing up. Physicians can now prescribe IL-ıRa to patients with rheumatoid arthritis and other illnesses, blocking IL-ı having become part of the therapeutic options to treat diseases like gout and many of the autoinflammatory disorders we are about to get to know.[32]

"The discovery of IL-1Ra opened up an entirely new field of human biology. One can say that it was a revolution in term of anti-cytokine therapy," Dr. Arend summed up. "My scientific career has been stimulated by an innate curiosity into basic biological mechanisms and their role in disease. I did not start out to discover a new therapeutic agent for human disease, but to pursue my curiosity. I have benefited by early exposure to an excellent mentor, collaboration with many superb scientists and research fellows, and the assistance of highly capable professional research assistants. The discovery of IL-1Ra followed a logical pathway in my research and has opened up entirely unanticipated benefits to the scientific world and to many patients."

"The discovery of IL-Ra was due to our frustration to lack of detection of IL-1's biological activity," concluded Dr. Dayer. Coping with that frustration by recognizing that "also the absence of reaction is a reaction" allowed scientists to successfully identify the origin of that mysterious absence.

A Distant Relative

As with IL-1 and IL-1Ra, researchers discovered the next family member, IL-18, through a function-to-structure approach. However, unlike the familiar issues of fever and arthritis that had steered investigators toward the first family members, the biological changes scientists were trying to understand when they discovered IL-18 were more recondite, somewhat disconnected from the appreciation of those not working in the immediate field.

Because they are numbered sequentially, sixteen interleukins (and many more cytokines that did not make it into the interleukin nomenclature) had been discovered in the eleven years that passed from cloning of IL-1 in 1984 to identification of IL-18 in 1995. During these years, investigators had started to chase body messages that were less directly connected to the subjective experience of illness, looking for mediators that exert their effects down the line, somewhere inside the intricate molecular network whose emergent activity generates the signs and symptoms of disease. To get to these once-removed messages, scientists needed to trick biology into revealing its deeper secrets by using experimental systems that may appear arcane and overly contrived.

Interferon-γ-Inducing Factor

Founded at the beginning of the 1970s, Hyogo College of Medicine in Nishinomiya, near Osaka, Japan, became the hub of research on IL-18, attracting many investigators working in the field. Dr. Hiroko Tsutsui was one of them, joining the college as a lecturer in the 1990s and rising to the rank of professor in 2006. Her doctoral thesis had focused on an experimental model in which liver macrophages (the same Kupffer cells Charles Dinarello had used during his early studies on IL-1) released a still-unidentified message when stimulated under very precise—and rather artificial—conditions.

Born in Osaka, Hiroko Tsutsui did not have any relative associated with science or medicine. In her junior high school days, she liked spending time after classes catching microorganisms from the school pond and examining them under a microscope. She also enjoyed preparing weather charts using information from radio broadcasts and making good use of her ham radio license. In winter, she liked to watch stars with a telescope. In high school she had become interested in mathematics, physics, and chemistry, but had found biology boring, as it did not seem to follow clear principles or rules. "It was like that until the day my biology teacher, Haga, enthusiastically described how nuclei contain DNA that determines the features and destiny of cells," Dr. Tsutsui wrote in response to my questions.

The teacher described how a three-letter code translates the ACTG language of DNA into that of amino acids. This type of biology, grounded in clear rules, fascinated me. However, my interests in social science and economics prevailed and thus I enrolled in the Faculty of Politics and Economics at Waseda University. But here, again, I found classes not meeting my deep interests. In particular, I was not satisfied with the economics of Keynes and Samuelson. Thus, remembering my earlier fascination for the genetic code, I decided to switch to life sciences. In Japan back then (as in many other places in the world), women with a PhD in basic biology rarely managed to build a significant career, whereas getting a job was easier with a medical degree. Therefore, I decided to study medicine. At that point I was already married. My husband, a science teacher in high school, had plans to obtain a PhD in economics, while I enrolled in premedical classes at Osaka City University.

When the time came to join a laboratory to obtain hands-on research experience, Dr. Tsutsui chose hepatology, the study of the liver. "I was assigned to do a repetitive task. Every day I was asked to do the same work, but I remember that I was very happy," she told me. "This experience made me decide to continue my career with a medical internship in hepatology while also pursuing a PhD in biochemistry, attempting to bring together clinical skills with strong basic research." That basic research shortly thereafter brought her to Hyogo College of Medicine and to IL-18.[33]

Just like IL-1, IL-18 was born with a different name, that of interferon-γ-inducing factor, or IGIF. It was so designated due to its ability to upregulate production of another cytokine, interferon-γ, in mice that had first been injected with the bacterium *P. acnes* and then with the bacterial product endotoxin. A commensal, *P. acnes* is a regular inhabitant of our skin but also participates in the pathogenesis of acne, as its name clearly indicates. However, the scientists who discovered IL-18 did not use *P. acnes* because they were trying to learn something about acne, or about any of the other diseases of which this bacterium might partake. These investigators were instead using *P. acnes* as a research tool aimed at laying open the pathways they wanted to understand. Immunologists had appreciated the inflammatory and immunostimulatory properties of *P. acnes* for decades. They knew, since at least the 1960s, that *P. acnes* strongly stimulated macrophages to synthesize messages that, in turn, activated lymphocytes and other leukocytes, pushing them to synthesize additional cytokines, primarily interferon-γ.

After going through its own long and remarkable process of discovery, interferon-γ was purified and cloned in the early 1980s, one of only three genes cloned before IL-1. The biological importance of interferon-γ cannot be overstated: it is a central regulator of immune responses, essential in defense against infections and cancer, while also being involved in the pathogenesis of many disorders. Among its many other actions, once synthesized and released by lymphocytes and other cells, interferon-γ loops back to stimulate macrophages (which had originally induced its production), upregulating these cells' ability to kill certain types of microbes, such as the mycobacterium that causes tuberculosis.

We can therefore state that interferon-γ is directly linked to the body's ability to clear infections, something we can classify as a first-level message. Those mediators that macrophages produce to push lymphocytes to make interferon-γ are instead one step removed from direct control of the infection, behaving as second-level messages whose job is not to act but to summon help. This is not to say they are not themselves crucial, as shown by the dramatically increased susceptibility to tuberculosis and other infections caused by mutations in genes coding for both first- and second-level messages.[34]

In the late 1980s, scientists discovered the cytokine IL-12 while searching for just such second-level mediators necessary for production of interferon-γ, that is, while looking for macrophage-derived factors that stimulated synthesis of interferon-γ by lymphocytes. However, some pieces of the puzzle were still missing, as IL-12 could not explain the whole story. The scientists at Hyogo College kept looking, with the goal of finding additional second-level mediators produced by macrophages and involved in induction of interferon-γ. To put themselves in the best possible situation, the researchers decided to use a system in which they already knew macrophage-derived factors would induce tremendous amounts of interferon-γ, implying that macrophages were likely producing a whole lot of the message the investigators were trying to identify. The system chosen was the one Dr. Tsutsui had used in her thesis: inject mice with *P. acnes*, wait a few days, and then inject them again with endotoxin. Instead of helping to clear an infection, the high levels of interferon-γ produced under these elaborate experimental conditions instead cause massive liver damage, as Dr. Tsutsui had contributed to demonstrate with her doctoral work. A potentially good response turned on its head, something that often happens with cytokines.

Being a hepatologist, Dr. Tsutsui was studying the *P. acnes*/endotoxin model with an eye to its effects on the liver, while immunologist Haruki Okamura was using the same system to search for the yet-to-be-discovered interferon-γ-inducing factor. That search was successful, marking the birth of IGIF, which was later renamed IL-18. Together with the coauthors of their 1995 publication in *Nature*, Drs. Okamura and Tsutsui purified and cloned IL-18 from mouse livers, analyzed regulation of its production, generated recombinant IL-18 to show its biological function, and also used subtraction approaches to demonstrate that IL-18 was indeed required for development of liver damage in the *P. acnes*/endotoxin model.[35]

This was a great achievement, but why should somebody care about which cytokine is responsible for inducing another cytokine in such an artificial experimental model that bears no resemblance to any human disorder? Because this is nothing but the first essential step in understanding whether the newly discovered body message might play an important role in disease and whether one could—once enough evidence would be obtained—possibly manipulate this cytokine's production or activity to alleviate human suffering. It took an arcane experimental setup to unveil the existence of a new body message so that scientists could then analyze its involvement in more familiar and relevant contexts. In fact, the discovery of IL-18 opened the doors to a flood of studies aimed at understanding the role of this cytokine in a variety of common disorders.

Soon after cloning of IL-18, its receptor was identified and the signaling pathways elucidated using *in vitro* techniques. Researchers generated mice deficient for IL-18 and its receptor, as well as transgenic mice overproducing IL-18 in different tissues, allowing investigation of the involvement of IL-18 in experimental models of human disease through powerful *in vivo* subtraction and addition approaches. Using the cornucopia of research tools that rapidly became available, Drs. Tsutsui, Okamura, and their colleagues at Hyogo College of Medicine— particularly Drs. Kenji Nakanishi, Tomohiro Yoshimoto, and Shizuo Akira—went on to study the role of IL-18 under several experimental settings. Other scientists from all over the world joined the quest, including our team directed by Dr. Dinarello at the University of Colorado. Studies demonstrated that IL-18 cooperates with IL-12 for production of interferon-γ, but that it also exerts a wide range of additional pro-inflammatory activities, participating in the pathogenesis of many diseases, from those involving the liver that Dr. Tsutsui had initially investigated, to inflammation of the skin and the colon, cardiovascular and metabolic disorders, cancer, allergies, and many more.[36]

Should We Call It Interleukin-1γ?

Immediately after publication of the article reporting cloning of IL-18, investigators at DNAX Research Institute in Palo Alto, California, used computer modeling algorithms to analyze the genetic and amino acid sequence of the new cytokine. They demonstrated close resemblance of the tridimensional structure of IL-18 (then still called IGIF) with that of IL-1α, IL-1β, and IL1Ra. In a one-page letter to *Nature*, the DNAX investigators proposed that IGIF was likely a member of the IL-1 family and therefore suggested the new cytokine be named IL-1γ.[37]

The name IL-1γ did not stick, but the authors were right on target about the new cytokine's structure, particularly about the relatedness of IL-18 with IL-1β. This observation jump-started the notion that the IL-1 family might be far more numerous that initially believed. Indeed, seven hidden relatives were soon discovered by recognizing their physical resemblance to the family members already known.

The Kids in Their Thirties

Each of the next seven members of the IL-1 family—IL-33, the four versions of IL-36, IL-37, and IL-38—was identified through structure-first methods that would not have been possible without the tools of bioinformatics. These discoveries clearly demonstrated that structure can be as powerful as function in finding new body messages, although deciphering the biological changes induced by these younger IL-1 family members proved to be at least as challenging as unveiling the structure of the older relatives, if not more difficult.

Going in Silico

Bioinformatics is a general term that refers to the use of computer programming to solve biological problems that are so complex as to require the support of artificial intelligence. This approach is called *in silico* to indicate that computer chips take the place of the more traditionally used molecules and cells (*in vitro*) or animals and humans (*in vivo*). Bioinformatics has become an essential tool in making sense of the avalanche of data obtained by neuroscience and microbiome projects that aim to understand the intricacies of our neuronal connections and the complexity of the microbial world that shares our body space. However, its most familiar and most common use has been in genetics, where powerful computers are employed to sort through the billions of bases of our DNA.

An apparently senseless sequence of just four letters, genes actually have an elaborate structure. Just as the elements of poetry—stanzas, rhymes, and the like—permit classification of poems into related genres, genes with similar structures can be arranged into families. Location is also important, since genes with similar functions often (though not always) sit near each other in the same section of a chromosome. When genes with analogous functions are physically located near each other (and sometimes even when they are not), it is likely they have come into existence through duplication of an original ancestral gene: related versions are then called homologs. Because homologs arise by duplication of a piece of DNA, they tend to retain at least some degree of sequence similarity with the original gene. Bioinformatics can take into account each of these aspects—gene structure, physical proximity, sequence similarity—and many more to come up with potential assembly of genes into families, and with other useful information.

Another branch of bioinformatics, called structural bioinformatics, is instead concerned with the shape of biomolecules, particularly proteins. Proteins are chains of amino acids, whose sequence is determined by the DNA code of their respective genes. Once you have that code, you can deduce the amino acid sequence. However, the twenty or so amino acids that make up proteins have an assortment of chemical properties. While some have a neutral electric charge, others are positive or negative, which makes amino acids attract or repulse each other just like the poles of a magnet. Moreover, certain amino acids prefer a water-based environment, others shy away from water and instead like to bathe in fat. The result of this diversity is that as soon as it emerges from the ribosomal factory, the amino acid chain spontaneously folds into a characteristic tridimensional shape that is essential in determining that protein's unique function. If you unfold the protein into a two-dimensional string (a process called denaturation), its function is forever lost. Structural bioinformatics looks at a protein's tridimensional shape with the goal of either understanding the protein's specific function, finding potential ligands (which can be natural, such as body messages, or artificial, such as new drugs), or analyzing similarities among proteins and classifying them into structurally related families.

Complementary to each other, the genetics and structural bioinformatics approaches have both been instrumental in identifying the seven new members of the IL-1 family.

A Group of Six

Between 1999 and 2001, eight articles by seven different research teams reported finding six potential new members of the IL-1 family using *in silico* approaches. Spurring this frenzy was the short 1996 letter to *Nature* that had proposed that IGIF (the future IL-18) should be named IL-1γ based on its tridimensional similarity with members of the IL-1 family.

It is remarkable for a one-page report like that *Nature* letter to have such a profound impact. A look at the affiliation of the authors of the eight articles reporting the discovery of new IL-1 relatives is instructive in understanding the phenomenon. Of the seven teams involved in the bioinformatics search for new cytokines, only one (at the University of Sheffield in England) was housed in academia, the traditional place of biomedical research. Each of the other seven articles originated in either a biotech, a pharmaceutical company, or a research institute funded by a company, all located on either the East or West Coast of the United States.[38]

Why this sudden corporate interest in finding new cytokines? It was biotech boom time, with sequencing of the human genome about to be completed, even earlier than expected. Some of the companies interested in finding new IL-1 relatives were active participants of that sequencing enterprise, while others had finally succeeded in bringing biotech-made drugs to clinical use. The year 1998, in fact, saw the approval of Enbrel for treatment of rheumatoid arthritis, a medication that works by blocking the activity of TNF, a pro-inflammatory cytokine we will meet in the next two chapters. Enbrel's entry into the market—together with the first promising reports of trials showing effectiveness of blocking IL-1 with IL-1Ra in rheumatoid arthritis—had invigorated the industry after the previous stumblings resulting from the sepsis trials mentioned a few pages back.[39]

Although they all went *in silico*, the seven groups used somewhat different strategies to discover the six new IL-1 family members, mostly mixing genetics with structural bioinformatics. The newcomer cytokines initially received a confusing array of IL-1-related names that were later changed to more standard nomenclature. They are currently called IL36α, IL-36β, IL-36γ, IL36Ra, IL-37, and IL-38.

However, the road to understanding the biology of these six body messages has been painfully slow and mostly fallen back into the hands of academic researchers. Part of the reason for such slow progress might reside in an apparent reticence in approaching the study of these mediators. Whereas thousands of articles on IL-1 were published in the fifteen years that followed its cloning (not to mention all those that came before cloning), less than two hundred publications had reported original research findings about the six new cytokines (all combined) fifteen years after their discovery, despite the exponential growth of publications in the area of cytokine biology between the two periods.

Initial evidence about the likelihood that IL-36, IL-37, and IL-38 would have a highly complex biology—multiple variants, convoluted compartmentalization, differences among species—might have caused some wariness in approaching them. But most scientists would (should?) consider such complexity as a dare to action rather than an impediment. Another, perhaps more important reason may reside in the skewed perception of the relative biological (un)importance of these cytokines due to the strategy by which they were discovered. The fact that these mediators were born as structures dissociated from a specific function was possibly perceived as a clue that these molecules might carry rather trivial messages that could safely be ignored, at least compared to mediators discovered based on their biological activity. In other words, if these six new cytokines played such an important role in the body's dynamics, why were they not discovered by going after their conspicuous function? That perception, if ever it was present, turned out to be unwarranted, at least for IL-36. In fact, seminal studies demonstrated that a mutation that reduces the function of IL-36Ra, the equivalent of IL-1Ra for IL-36 cytokines, causes generalized pustular psoriasis, an extremely rare, life-threatening condition that had remained highly mysterious for a long time.[40]

Understanding the origins of disease is always important, no matter how infrequent the condition might be. However, the extreme manifestations of rare diseases such as generalized pustular psoriasis have often been instrumental in unlocking biological pathways involved in much more common conditions, conceptually playing the same role as the arcane experimental system that allowed researchers to identify IL-18. Following up on the implication of IL-36 in the rare and severe form of psoriasis, scientists in fact discovered that members of the IL-36 family are involved in the pathogenesis of the common form of that disease, a more limited autoimmune skin disorder, and likely in many other conditions as well.[41]

That the six new cytokines were discovered with little to no background research that might even hint at their possible function is possibly a third and related reason to explain scientists' lackluster interest. This paucity of prior research left investigators without a toehold to grasp where to even begin their studies, with some even questioning whether these molecules should be thought of as interleukins at all. But slowly, piece by piece, evidence is being assembled, showing that IL-37 has anti-inflammatory properties and that IL38 might have them too.[42]

Despite having been discovered using a structure-first approach like her six siblings, the last member of the IL-1 family, IL-33, was born with enough of a legacy to avoid most of these issues.

Carry-on Baggage

Skirting the rules of sequentiality, scientists discovered IL-33 five years after identification of IL-36, IL-37, and IL-38, but IL-33 received its current designation before its cousins, hence the lower figure in its name. The path that led to identification of IL-33 is quite peculiar, as was the place where this body message was discovered.

The research institute/biotech DNAX, home of the investigators that, based on structural analysis, had proposed classifying IL-18 as part of the IL-1 family, was founded in 1980 in Palo Alto, California, by three prominent researchers from nearby Stanford University: Arthur Kornberg, recipient of the 1959 Nobel Prize for his discovery of mechanisms involved in the synthesis of DNA; Paul Berg, who received the Nobel Prize for his studies on the biochemistry of nucleic acids the same year DNAX was founded; and Charles Yanofsky, selected for several prestigious recognitions, including the 1971 Lasker Award for his contribution to molecular genetics.[43]

In 1982, the pharmaceutical company Schering-Plough acquired DNAX, which, however, managed to maintain an unusual level of independence. "DNAX was an extraordinary place," according to Dr. Maria Grazia Roncarolo, who worked there for nine years (she was not part of IL-33's discovery, but we will meet her again in the next chapter). "The reason why DNAX was able to perform in a very academic way is that the three professors who founded it had tied selling the institute to Schering-Plough with the binding that they would remain in the board to define strategies for ten years." The team of scientists at DNAX included immunologists, molecular biologists, and bioinformaticians, who combined their complementary expertise to come up with new ways to identify cytokines, including evaluation of structural similarities. On top of multiple additional crucial findings, DNAX scientists discovered an impressive number of cytokines: IL-4, IL-10, IL-23, IL-25, IL-27, and IL-33.[44]

With the exception of IL-4 and IL-10, each of the other cytokines identified at DNAX was discovered through a structure-first approach. Dr. J. Fernando Bazan, an expert in structural bioinformatics and the first author of the letter to *Nature* that analyzed IL-18's structure, was an essential member of the teams that discovered IL-23, IL-27, and IL-33. "I remember Fernando coming to us in the human immunology group and saying, 'This could be a cytokine, this could be another one,' all this through computer designer predictions. He is really a genius. The molecular biologists would generate short sequences, small pieces, and on these small sequences he did all the mimicry. It was really a great time," Dr. Roncarolo continued.

After high school in Costa Rica, Fernando Bazan obtained a BS in physics at Stanford University, followed by a PhD in biophysics at the University of California, Berkeley. Even though he was studying at Berkeley, Bazan spent a great deal of time across the bay at the University of California, San Francisco, learning the principles and practice of protein folding and bioinformatics. After graduation, he formally moved to the University of California, San Francisco, focusing on viral proteins but beginning his involvement with cytokines and their receptors. He joined DNAX in 1994 and remained there until 2002, when he moved to another Bay Area biotech, Genentech.[45]

It was during Dr. Bazan's last few years at DNAX that IL-33 arrived on his plate. In the late 1980s, various research teams had discovered a class of molecules related to the IL-1 receptor. These had, however, remained orphans, that is, receptors without a known ligand. One of these orphans, known as ST2, appeared to be involved in the pathogenesis of allergies and asthma, thus generating quite a lot of interest as a potential therapeutic target for these conditions. By peering into databases and then confirming their findings with more classical bench experiments, Drs. Bazan, Robert Kastelein, and colleagues at DNAX identified IL-33 as the ligand of the no-longer orphan receptor ST2. Their 2005 publication in the journal *Immunity* covered the whole spectrum of approaches, beginning with fishing out IL-33 *in silico* using computational tools, then moving *in vitro* to analyze production and activity of IL-33 in cultured cells, and finally using *in vivo* addition systems to investigate the effect of recombinant IL-33 in mice.[46]

Thus, at variance with the other members of the IL-1 family identified *in silico*, IL-33 joined the world of cytokines by carrying its own baggage of prior research on ST2, arriving with at least a partial function already attached to its structure. Researchers in fact knew that ST2—the receptor for IL-33 and the prime mover for its discovery—was involved in the pathway leading to allergic disorders. Indeed, they went searching for IL-33 precisely because they were interested in understanding allergy and asthma. This made IL-33 a less intimidating object of research than its six siblings discovered *in silico*, since previous knowledge obtained by studying ST2 could be used as a scaffold to further investigate IL-33's biology. By itself, IL-33 in fact generated at least five times more publications than all of the other six new interleukins combined, even though the discovery of IL-33 is five years younger than they are.

Beginning with the launching pad of allergy and asthma, investigators expanded their knowledge of the biology of IL-33 to roles in tissue regulation, organ injury and repair, cancer, and immune responses. However, the precise behavior of IL-33 turned out to be extremely complex, somewhat similar to that of its cousin IL-1α. Both cytokines are unusual in that they can act as regular body messages outside of cells, but also have key functions in the intracellular space, even in the nucleus itself, making them tough cookies to understand. Another issue that has beleaguered scientists is whether IL-33, like some of its relatives, needs to be further handled—to reach maturation—before its message can convey its full meaning.[47]

The Family Handlers

Body messages with potent bioactivity are often synthesized as precursor proteins that must be cut into two or more segments before they can become fully active. An example is insulin, from which pieces have to be removed before this hormone can accomplish its tasks. This maturation, or processing, step probably represents a way to control the activity of mediators that are very powerful and could be damaging to the organism should their production be even slightly unregulated.

Some of the pro-inflammatory members of the IL-1 family become fully active only after they have matured. These cytokines deliver such powerful messages that many checks and balances are necessary to rein them in. Specific antagonists, like IL-1Ra and other molecules, can modulate these cytokines' activity, but processing of the amino acid chain—maturation of the protein—represents an additional checkpoint that takes place even before the cytokines are released from their manufacturing cells.

That IL-1α and IL-1β are first generated as precursor proteins that need to be shortened before they can bind to and activate their receptor was already understood at the time of their discovery. In fact, evidence indicated that the genes for IL-1α and IL-1β code for proteins larger than the ones scientists were detecting in cell culture supernatants. It looked like cells started out by making large, precursor IL-1α and IL-1β and then a segment was cut off from one end of the amino acid chain before the cytokines were released from the factory cell. Studies directed at understanding the mechanisms of these cytokines' processing— particularly those looking at maturation of IL-1β—not only answered the original question by identifying the maturation mechanisms, but also opened the road to momentous discoveries in seemingly unrelated areas.[48]

Interleukin-1β-Converting Enzyme

Enzymes are proteins that modify molecules.[49] Almost nothing would happen in our body without enzymes. We would not be able to digest food (proteases break down proteins into amino acids, amylase takes apart starch, lipase works on fat, lactase—for those who make it—on the sugar in milk), we could not absorb nutrients (we need enzymes in our intestine to do that) or regulate their distribution in the body (enzymes in the liver are necessary to accomplish this). Our muscles and heart wouldn't contract, our brain would be silent, our cells dead.

It is enzymes that process precursor cytokines; since cytokines are proteins, these enzymes belong to the general category of proteases. When time comes for the cytokine to mature, a specific protease attaches to the cytokine's amino acid chain and makes a very precise cut, always in the same place, between the very same two amino acids in the chain. This is critical, because a cut in any other position would not allow the mature protein to assume its final shape, essential for binding to and activating its receptor and therefore delivering its message.

The enzyme that processes IL-1β—initially named interleukin-1β-converting enzyme—was discovered in 1989 and cloned in 1992 by researcher at two companies that were racing against each other, Immunex and Merck. Here was a new enzyme, unrelated to any known molecule, that could represent a novel target for development of therapeutic agents.[50]

Many drugs work by blocking enzymes' activity. Aspirin inhibits the enzyme cyclooxygenase, which generates the inflammatory mediators prostaglandins, statins block an enzyme known as HMG-CoA reductase that is important in production of cholesterol, some drugs that control blood pressure inhibit other kinds of enzymes, and so on. Small chemical compounds that can be manufactured in a relatively inexpensive way by the chemical laboratories of pharmaceutical companies can generally pull off the job of blocking the activity of an enzyme. In addition, these are often drugs that people can take orally, in a pill, capsule, drops, or syrup. Compare these characteristics to those of a protein-based drug such as IL-1Ra, which was being developed by Synergen as the discoveries we are discussing here were taking place. Proteins are cumbersome and very expensive to produce; technicians need to grow vast amounts of recombinant bacteria in huge vats, as mentioned a few pages back. Furthermore, the stomach's digestive enzymes destroy protein-based drugs, making the convenient oral route of delivery unavailable. The only choice is to administer these compounds through injections, with all the issues that concern the use of needles. Clearly, targeting IL-1 by blocking an enzyme using small nonprotein compounds would be a much more attractive strategy than having to use a recombinant protein like IL-1Ra. This is likely one of the chief reasons that brought a pharmaceutical company like Merck into the race to identify interleukin-1β-converting enzyme. Small chemical compounds that block enzymes' activity are in fact the traditional arena of pharmaceutical companies, whereas biotechs like Immunex, Synergen and Amgen mostly deal with protein-based therapeutics (and, more recently, other biological molecules). Indeed, reports of small molecules able to inhibit the activity of interleukin-1β-converting enzyme appeared in the scientific literature soon after this enzyme's identification. Studies demonstrated that these inhibitors could block release of IL-1β from cells, therefore suppressing this cytokine's ability to cause inflammation and thus reducing disease severity in many experimental models of disease.[51]

While the discovery of interleukin-1β-converting enzyme was taking place, scientists at Hyogo College were in the process of identifying IL-18. In their 1996 letter to *Nature*, Dr. Bazan and his DNAX colleagues had suggested that IL-18 might be a member of the IL-1 family not only based on IL-18's tridimensional structure, as mentioned above, but also because the investigators had found a motif in IL-18's amino acid chain that corresponded to the precise site where interleukin-1β-converting enzyme cut IL1β. Indeed, in less than a year, two articles formally demonstrated that IL-18 underwent the very same maturation process as IL-1β and that the chopper was nothing but interleukin-1β-converting enzyme.[52]

The enzyme's name had therefore become inadequate, since not only IL-1β but also IL-18 could be counted among its targets. This turned out not to be an issue, though, since a brand-new name had just been bestowed on interleukin-1β-converting enzyme by a completely separate group of researchers. This enzyme's uniqueness had in fact attracted the attention of investigators from a different field, leading to an identity change for interleukin-1β-converting enzyme and to the launch of a parallel, highly consequential, series of rapid discoveries.

Caspase-1

Apoptosis, also called programmed cell death, is one of the ways in which cells can die. It is a very orderly process, initiated in response to either internal or external cues. Almost every facet of life requires this type of death, particularly in multicellular organisms like humans. Apoptosis is involved in embryogenesis, in shaping the nervous system, in the proper functioning of the immune system, in the turnover of cells in the gut and other organs, and in uncountable additional processes.[53]

Enzymes keep cells alive by catalyzing a myriad of biochemical reactions but, perhaps counterintuitively, enzymes are also necessary for cells to die peacefully by programmed cell death. Several of the enzymatic pathways that regulate apoptosis were first discovered in the tiny worm *C. elegans*—the preferred model organism for investigation in this field—and their corresponding genes and proteins later identified in humans. In 1993, that process was reversed.

First discovered in mouse and human cells by investigators studying IL-1β processing, the equivalent of interleukin-1β-converting enzyme was later identified in *C. elegans* by scientists working in the field of apoptosis. They named the worm version ced-3 and demonstrated its fundamental role in programmed cell death. A rapid cascade of discoveries followed, so that in just three years ten distinct genes related to interleukin-1β-converting enzyme had been discovered in humans and other organisms, all involved in the process of apoptosis.[54] Scientists assigned a bewildering array of names to the ten new enzymes, and it became rather difficult to keep track of what was what. To put some reason into the nomenclature, in 1996 the term *caspase* was introduced. The new name indicated the unique manner by which this family of enzymes cuts its substrates: *c* for its dependence on a cystein amino acid for cutting, *asp* for the aspartic acid amino acid after which the enzyme usually cuts its substrate, *-ase* as the common suffix for enzymes. Interleukin-1β-converting enzyme thus began a new life as caspase-1.

With these discoveries, the field of apoptosis, already a highly active area of investigation, expanded even further and its connections to inflammation were strengthened. Dr. H. Robert Horvitz, who had led the 1993 study reporting discovery of caspase1 in *C. elegans* (as well as many prior and subsequent seminal studies), shared the 2002 Nobel Prize with Sydney Brenner and John E. Sulston for discoveries in programmed cell death.[55]

Determination of the role of caspases in apoptosis in turn contributed to paving the way for a second major achievement directly derived from the study of IL-1β and IL-18 processing.

Inflammasomes

After identification of interleukin-1β-converting enzyme/caspase-1, a nagging issue remained. Investigators had demonstrated that caspase-1, the processing enzyme, itself needed to be processed—cut into two parts—before it could act on IL-1β and IL-18. This made biological sense: if generation of active IL-1β and IL-18 needed to be tightly regulated, you couldn't have their processing enzyme constantly turned on. However, how this processing of the processor might occur was not understood. All that was known, for a long time, was that efficient release of active IL-1β from cells required two stimuli: one to induce production of the cytokine, the other to activate its maturation and release.[56]

A cluster of observations made by immunologists dealing with rare diseases, by experts in apoptosis, and by biochemists interested in understanding how cells sense the presence of danger eventually converged and solved the problem with the discovery of a new macromolecular entity, the inflammasome.

Autoinflammatory syndromes are rare diseases characterized by recurrent attacks of inflammation. Individuals affected by a variant of these diseases, familial cold autoinflammatory syndrome, develop fever, skin rashes, and other troublesome problems that persist for a day or so after exposure to cold. Another form, Muckle-Wells syndrome, leads to periodic assaults of abdominal pain, arthritis, and skin rashes arising from seemingly nowhere. Other variants of autoinflammatory diseases come with their own set of symptoms, but all involve inflammation that is apparently unprovoked or is precipitated by situations that are harmless to most people, such as cold. These diseases had been recognized as distinct pathological entities since the 1940s, but it was only in the late 1990s that investigators began identifying their causative mutations.

The first finding came in 1997, when a French consortium of scientists determined the mutation that causes the autoinflammatory syndrome familial Mediterranean fever. The gene mutated in individuals affected by this syndrome encoded a protein of unusual structure and of unknown function, which the investigators alternatively named marenostrin from the Latin name of the Mediterranean Sea (*mare nostrum*, our sea) and pyrin, since it was involved in periodic fevers.[57]

Meanwhile—it's now 1999—researchers studying apoptosis were discovering a new family of proteins related to both caspases and the newly described marenostrin/pyrin. At the same time, a separate line of investigation by biochemists found those same molecules participating in sensing of bacteria by immune cells. Nomenclature in this area was (and still is) so confusing that it's better not to even begin naming names.

Results from these three apparently unrelated fields—autoinflammatory syndromes, apoptosis, and bacterial sensing—converged in the early 2000s. First, Dr. Hal Hoffman and his colleagues at the University of California, San Diego, discovered that mutations in a gene currently called NLRP3 caused familial cold autoinflammatory syndrome and Muckle-Wells syndrome. The protein synthesized by this gene belonged to the pyrin family, the same family as the proteins that caused familial Mediterranean fever, but its specific function, like that of marenostrin/pyrin, remained unknown.

A few months later, the team of Jürg Tschopp at the University of Lausanne, Switzerland, coined the term *inflammasome* to describe the orderly assembly of a group of proteins that join together in response to bacterial sensing, activate caspase-1 by splitting it in two, and therefore cause maturation of IL-1β and IL-18. These investigators, and others, demonstrated that a close relative of NLRP3 (the gene Dr. Hoffman had found mutated in patients with autoinflammatory syndromes) was a central element of the inflammasome, connecting bacterial sensing to generation of mature IL-1β and IL-18. Soon scientists demonstrated that NLRP3 itself was directly involved in formation of the inflammasome, thus linking danger-sensing structures to autoinflammatory syndromes. Additional caspase- and pyrin-related proteins discovered by apoptosis investigators were soon also brought into the circle of inflammasome components, putting together all the pieces. Our current understanding of the process goes something like this: when a cell senses danger, various pyrin-related proteins, including NLRP3 and its relatives, assemble to form a macromolecular structure—the inflammasome—that processes caspase-1, which then cuts IL-1β and IL-18, turning them into mature and active cytokines that can be released from their producing cell and deliver their inflammatory messages to target cells. In patients with certain forms of autoinflammatory diseases, genetic mutations cause parts of this mechanism to be either constantly active or triggered by otherwise innocuous stimuli, such as exposure to cold, generating an array of symptoms that resemble a nonexistant infection or autoimmune response.[58]

Several types of inflammasomes have since been discovered. They contain various kinds of proteins, but all have a comparable structure and function. Inflammasomes help cells sense the presence of danger, which can take the form of microbes but also assume many other shapes, such as metabolic disruption or tissue damage. Consequently, inflammasomes have been implicated in the pathogenesis of infections, diabetes, obesity, atherosclerosis, gout, neurodegenerative and autoimmune disorders, cancer. You name it, there is probably not a single type of disease that has not been linked to the inflammasomes. Although the pathways that activate inflammasomes are diverse, the downstream outcome is always the same (at least as far as we currently understand it): processing of caspase-1 and, therefore, maturation of IL-1β and IL-18, with consequent induction of inflammation, the common pathogenetic pathway that links all the disparate conditions in which inflammasomes have been involved.[59]

The extreme and peculiar phenotypes of rare autoinflammatory syndromes, such as familial cold autoinflammatory syndrome, Muckle-Wells syndrome, and familial Mediterranean fever, served as the initial window to peer into pathways that eventually deepened understanding of much more common conditions, just as we described for IL-36Ra and psoriasis. However, individuals with autoinflammatory syndromes have not simply been helpless experimental subjects, providing information that might help others but receiving no personal benefit. Knowledge that autoinflammatory syndromes result (for the most part) from spontaneous activation of caspase-1 and excessive production of IL-1β and IL-18 provided a rationale for treatment of these illnesses. Administration of IL-1Ra has indeed become the standard of therapy for patients with some types of autoinflammatory syndromes, highly improving their quality of life, while blocking IL-18 is also beginning to emerge as an effective therapeutic strategy.[60]

Back to Fever

With autoinflammatory syndromes, we have come all the way back to fever, so let's also return to Tamas Bartfai, the neuroscientist who, at the beginning of this chapter, had told us that the evolutionary meaning of fever remains unclear. During our interview, Dr. Bartfai also briefly told me about major events in his life.

> I consider it an accident of life to have been born in Hungary, the offspring of a Jewish family that had been decimated by the Hungarian Arrow Cross and the German Nazis. However, I was able to benefit from Hungary's excellent educational opportunities in mathematics and chemistry. I profit from it daily, but I had to flee in 1971 after I was arrested several times for wanting a more humane system. I went to Sweden, where I obtained a PhD in biochemistry in 1973. During those first years in Sweden, as an orphan with no means of support, I had to somehow make a living. Accidents and accidents,

the laboratory where I started as a dish washer was in biochemistry, but it could have been in physics. I washed the dishes and discovered how poorly experimental design was applied even in a well-known laboratory. I ended up writing a thesis on mathematical models in biology that gained me a fellowship at Yale University and launched me into a lifetime of peripathetic research.[61]

Peripathetic indeed. In addition to Sweden, where he has been professor and chair in neuroscience and biochemistry at the University of Stockholm and the Karolinska Institutet, Dr. Bartfai worked in France, the United Kingdom, Switzerland, where he was senior vice president at Hoffmann-La Roche for five years, as well as in several institutions in the United States, including the Scripps Research Institute in La Jolla, California.[62]

Although he made influential contributions to various areas of neurobiology, fever has always been among Dr. Bartfai's favorite topics of investigation. "When I entered the field, I was annoyed that, while in sickbed, I had to take my temperature several times daily and yet doctors did not have—and incredibly we still do not have—much clue about the significance of fever. We did it because it was easy to measure, cheaply and quantitatively," Bartfai told me. "Fever was a great neurobiological topic for me, a central reflex that can be studied without psychological tests, quantitatively, and where my colleagues and I could move between molecules in the hypothalamus and *in vivo* measurements of heat generation and temperature."

The hypothalamus Dr. Bartfai mentions is the part of the brain that controls most of the body's basic functions, including temperature regulation. When leukocytes make IL-1 in response to microbial stimulation, or any other situation that causes fever, such as the autoinflammatory syndromes we just described, IL-1 binds to its receptors on target cells, which in turn activate the enzyme cyclooxygenase. This enzyme generates molecules, called prostaglandins, that act on neurons in the hypothalamus to reset the body's temperature settings, resulting in fever.[63] Prostaglandins are involved in many additional responses, including modulation of inflammation and pain. When we take an aspirin or similar drugs for a fever or a headache, it is cyclooxygenase, and therefore production of prostaglandins, we are trying to block. Evidence that prostaglandins are the molecules that cause fever emerged in 1971. In their 1972 review simply entitled *Fever*, Drs. Atkins and Phyllis Bodel already discussed the potential connections between endogenous pyrogen (one of IL-1's original names), prostaglandins, and aspirin-like drugs.[64]

Is IL-1 the only body message that can upregulate prostaglandins and cause fever, the only endogenous pyrogen? The answer is no: some of the pro-inflammatory cytokines we are going to meet soon, particularly TNF, which is just about to come up in the next chapter, can also induce fever. However, IL-1 remains one of the most potent pyretic messages.[65]

Even though we still don't understand what fever is good for, as Dr. Bartfai reminded us, thanks to the research we reviewed in this chapter, and many more investigations, we know a good deal about how fever comes about.

A Final Snapshot

Having circled all the way back to IL-1 and fever, one of the very first observations that sparked discovery of the IL-1 family of cytokines, let's take a final, quick look at the whole clan. We moved from the early attempts to identify endogenous pyrogen by measuring rabbits' body temperature with rectal thermometers, through cell cultures that refused to respond to IL-1, thus unveiling the presence of IL-1Ra, to the use of arcane experimental systems that were instrumental in the discovery of IL-18, all the way to the sophisticated bioinformatics tools that served to model the new generation of IL-1 family members. We peered into the broad avenues opened up by identification of caspase-1 and into the role rare genetic diseases play in the scientific process. We've also seen how new body messages can be discovered by chasing their function or by peeking at their structure, as well as how important prior findings are in shaping the direction of new projects. Lastly, this chapter introduced the concept that the activity of potent pro-inflammatory mediators needs to be tamed and controlled. We talked about body messages—IL-1Ra and IL-36Ra—that specifically inhibit the activity of their respective cytokines. The next chapter will expand this concept by examining the balance between pro- and anti-inflammatory mediators under a wider perspective.

4

Achieving Balance: Pro- and Anti-Inflammatory Mediators

"THERE ARE NO bad kids, only constipated ones," ran a famous Italian TV advertisement of my childhood, explaining to parents how just a spoonful of a sweet white powder could turn their cranky pests into delightful little angels. Thankfully, that sugary laxative is long defunct, though the ad's underlying concept is not. If you have never searched the World Wide Web for "intestinal balance," I highly suggest you try. The results are almost surreal, from diets named after hymns and prayers to the seven steps to fix your gut, on to countless pills, magic foods, and other less pleasant interventions that promise to realign, readjust, and reform you, so that you can resurface as the better human being you never knew you were.

Long before development of modern medicine, with a specific drug for each specific ailment, laxatives were among the few therapeutic tools available to physicians, widely prescribed for all sorts of conditions. Although laxatives' ability to cure may be questionable, their use certainly culminated in highly visible effects. Laxatives' modern versions, advertised as ways of removing toxins supposedly accumulated through unhealthy lifestyles and thus aimed at purifying both body and soul, are solidly entrenched in our culture, which assigns a strong moral connotation to regularity. The long history of purgatives as methods of punishment and public shame is also sadly familiar to many, and perhaps even more sadly forgotten by many more.[1]

In the last few years, the science of the gut microbiota, those billions of bacteria and other microbes that live inside our digestive system and are essential to our well-being, has rekindled fascination with the intestine as the source of all that is unwell with the body. While that science is solid, and getting better every day, its overinterpretation unfortunately runs rampant, with unscrupulous quacks peddling unproven remedies, just like their predecessors did for centuries.

Intestinal balance, though, is serious business for scientists. Some rigorously investigate the microbiota, others look at the lining of our bowels, while still others peer into the network of mediators that maintains the gut's integrity. In this chapter, we will address an assortment of body messages that strongly contribute to maintenance of gut homeostasis, of intestinal balance if you wish. However, none of the mediators we will meet—tumor necrosis factor, IL-10, chemokines, and annexin A1—was discovered anywhere near the intestine. Although scientists identified these molecules through a function-to-structure approach, a unique, distinct research process is behind each of them.

After reviewing the discovery and biology of these mediators one by one, we will bring them all together to examine their coordinated ballet in the depth of our gut. In the pages that follow, we will also take a good look at the lives of some of the researchers who discovered and initially studied these body messages.

Tumor Necrosis Factor

Just like the IL-1 family of cytokines that was the protagonist of the previous chapter, a separate class of mediators can be assembled into the tumor necrosis factor (TNF) family, which, at nineteen members, is even more numerous than the first family. Whereas in Chapter 3 we separately reviewed each IL-1 relative as a way to appreciate the evolution of body message discovery, here we focus instead on a single mediator, TNF, a cytokine that has reached popularity as a pharmacological target for treatment of chronic inflammatory diseases, including those of the intestine.

Despite its ostensible link to the death of cancer cells—after all, TNF stands for *tumor necrosis factor*—this cytokine is mostly studied for its potent pro-inflammatory effects. Similar to IL-1, leukocytes synthesize TNF in response to microbial stimuli and other situations of perceived danger, this cytokine acting as a central player in both induction and coordination of the inflammatory response that helps clear infections but, at times, provokes damage itself. However, TNF is also involved in tissue homeostasis, being an important determinant of cells' fate and thus participating, among other processes, in the delicate balance of renewal of our intestinal lining.

Again in similarity with IL-1, parallel paths of investigation led to identification of TNF. Because one of these paths runs through Kenya, a quick detour to this African country can get us started on our exploration of the discovery of TNF.

Skinny Cows

The protozoa *Trypanosoma* are unicellular organisms that generate much human and animal suffering. The insects popularly known as kissing bugs spread a type of trypanosome called *T. cruzi* from person to person, causing Chagas disease, a condition that affects millions of people and leads to thousands of deaths every year, mostly in South American countries. Across the Atlantic, in sub-Saharan Africa, it is instead tsetse flies that transmit another type of trypanosome, *T. brucei*, the agent of sleeping sickness, a terrible disease that also kills thousands of people annually. *Trypanosoma* also infect African cattle: cows become emaciated, eventually wasting away and dying, causing not only misery for the animals but also crippling economic consequences for the people whose livelihood depends on these cattle.

Cachexia is the term that defines the wasting syndrome of cattle with parasitic trypanosomiasis, a term that originates from the Greek words *kakos*, which means bad, and *hexis*, condition. Cachexia is a bad catabolic condition indeed, with the organism rapidly depleting its fat reserves and, even worse, breaking down massive amounts of muscle and its proteins. The resulting weight loss and weakness are remarkable and unmistakable. Cachexia is not only a problem of infected African cattle; a considerable percentage of patients with

AIDS, cancer (especially tumors of the gastrointestinal tract), and several other chronic diseases often die as a direct result of cachexia. Despite much research, we still do not understand clearly what causes cachexia and have no effective way of treating it.

Anthony Cerami is the scientist who went to Kenya to study trypanosomes and, in the process, wound up identifying the cytokine we now call TNF. After receiving a PhD from the Rockefeller University in New York City and obtaining research experience at both Harvard University and the Jackson Laboratory, in 1969 Dr. Cerami returned to the Rockefeller University as a faculty member. In the late 1970s, he received a ten-year grant from the Rockefeller Foundation to study parasitic diseases. That grant was the direct result of a program established by Kenneth Warren, a physician-scientist passionate about the fight against infectious diseases, particularly those affecting the poorest countries. In 1977, Dr. Warren had become director of Health Sciences at the Rockefeller Foundation, where he established the Great Neglected Diseases Network, meant to link "the bench with the bush".[2]

Dr. Cerami's grant focused on development of potential treatments for cachectic African cattle infected with trypanosomes. After studying a number of compounds in mice and rats in his New York City laboratory, Cerami traveled to Kenya to test the drugs directly in infected animals in the field. Although that line of investigation turned out to be unproductive, the test compounds mysteriously precipitating cows' demise instead of averting it, those studies led Cerami to question the origins of cachexia in response to trypanosomiasis. It was such an overwhelming response caused by a relatively small number of bugs! This line of thinking sparked a rather simple, yet brilliant, hypothesis: production of an endogenous mediator by the animal's own cells as a reaction to trypanosomes would cause the disproportionate cachectic response.[3]

Cerami soon returned to his laboratory and continued his experiments using rabbits instead of cows, trying to find out whether a soluble mediator—a body message—indeed caused trypanosome-induced cachexia. During the course of these studies, his research team accidentally discovered that the blood of cachectic rabbits was full of lipids, a thick layer of yellow fat floating on top of blood samples collected in test tubes. The investigators traced the molecular mechanism for this lipid accumulation to a metabolic alteration, that is, to inhibition of an enzyme called lipoprotein lipase that is responsible for uptake of lipids into adipose tissue. When the activity of lipoprotein lipase is reduced, as is the case in cachectic rabbits, adipose tissue cannot pick up lipids from blood, with fat therefore accumulating in the circulation.

Cerami and the postdoctoral fellow working with him, Dr. Masanobu Kawakami, now started to experiment in mice and also switched their experimental system from parasites to bacteria. In this way, they were able to demonstrate that the bacterial product endotoxin inhibited lipoprotein lipase in mice just as trypanosomes had done in rabbits (and possibly African cattle). Then, utilizing an approach similar to the one Drs. Atkins and Wood had employed in their research on endogenous pyrogen reviewed in the prior chapter, that is, by transferring serum between animals, Drs. Cerami and Kawakami proved that a substance made by the mouse's own cells was responsible for this metabolic effect. Just as had been for endogenous pyrogen, this result clearly signaled the presence of a body message, of a molecule made by the cells of the mouse in response to bacterial stimulation.

Following this insight was a series of experiments that again mirrored those that had led to identification of IL-1, with Dr. Cerami's team demonstrating that macrophages (the same leukocytes that make IL-1) also synthesized the new body message, and identifying many of the protein's biochemical characteristics. The researchers managed to partially purify the new mediator, but were unable to obtain its amino acid or genetic sequence (let's recall the technical difficulties in performing these tasks in the early 1980s). Undeterred by their lack of sequence, in 1981 Drs. Cerami and Kawakami nevertheless submitted an application to the United States Patent Office, naming their newly discovered mediator cachectin, for the original source of its discovery (cachectic animals). That 1981 patent described cachectin's general biochemical characteristics and methods of induction, but also outlined the principle that inhibiting cachectin might have therapeutic value in a wide range of diseases, a principle that would have important ramifications in the decades to come.[4]

Despite having a patent, Cerami still needed to figure out the exact nature of cachectin, its amino acid and genetic sequence. To this aim, he enlisted the help of Dr. Yu-Ching Pan, at the pharmaceutical company Hofmann-La Roche, who succeeded in obtaining part of cachectin's amino acid sequence on his very first attempt. At that point, Dr. Bruce Beutler, who had joined Cerami's team as a postdoctoral fellow, noticed a close similarity between the newly sequenced mouse cachectin and a human protein that other investigators had recently cloned and named tumor necrosis factor.[5]

Indeed, in 1984 another team had happened on the same body message by taking a completely different route, yet one that still began with microbes.

Killing Cancer

"History repeats, but science reverberates," writes Dr. Siddhartha Mukherjee in his epic history of cancer as he ponders the difficulties of predicting the direction of scientific advances and how individual discoveries may have utterly unexpected outcomes.[6] Indeed, when William B. Coley, a surgeon at Memorial Hospital in New York City, published an 1893 report describing the ability of a mixture of bacterial components to induce remission of certain types of cancers, he almost certainly did not expect his actions would eventually result in the discovery of a new body message that would become the target for treatment of a completely different category of diseases.[7]

Several clinicians and scientists followed up on Dr. Coley's observations of cancer remission in response to bacterial concoctions, but results were variable and side effects could be quite severe. One of the scientists who picked up research in this area was Dr. Lloyd Old, a California-born physician and accomplished classical violinist who made multiple seminal contributions to cancer research. By injecting mice and rabbits with a bacterial mixture similar to Dr. Coley's concoction, in a 1975 publication Drs. Elizabeth Carswell, Lloyd Old, and colleagues demonstrated production of a substance endowed with the ability to destroy cancer. Their approach was conceptually similar to those I already described: inject animals with bacterial products, wait an appropriate amount of time, take the animals' serum, and either inject it into a separate series of mice that have tumors under their skin or squirt it on top of cancer cells growing in a dish; watch cancer die. These scientists named the substance they discovered tumor necrosis factor (TNF), perhaps inspired by a similar term, tumor-necrotizing activity, introduced by Edward O'Malley and colleagues in a 1962 publication that reported the results of similar serum-transfer experiments. Several additional studies by Drs. Old, O'Malley, and other investigators demonstrated that activated macrophages produced TNF, again in stark parallel with the almost contemporaneous process of identification of IL-1. Debate about the actual pathophysiological role of TNF ran rampant: many investigators considered this mediator's ability to kill tumors of secondary relevance, proposing instead that its most important function would be in protection against infections. However, the precise chemical characteristics of this early version of TNF were never determined.[8]

Enter Dr. Bharat B. Aggarwal. Born and initially educated in India, Aggarwal had obtained a PhD in biochemistry at the University of California, Berkeley, and later spent a period as a post-doctoral fellow at the San Francisco campus in a laboratory focusing on reproductive hormones. In 1980, Aggarwal joined the recently founded biotech Genentech, in nearby South San Francisco. It wasn't long before Bob Swanson, Genentech's founder and president, "walked into Dr. Aggarwal's office and asked him to find a 'cure for cancer.' When Dr. Aggarwal asked how he was supposed to do that, the immediate answer of Bob was that lymphokines are a cure for cancer and if he could clone this 'sucker' he'll have a cure."[9] Because Dr. Aggarwal came from an unrelated scientific background, he did his homework and learned that the term lymphokines referred to protein mediators produced by activated lymphocytes.[10]

Armed with this and additional information, Dr. Aggarwal got to work together with his colleagues on finding the "sucker" that would cure cancer. They based their strategy on the use of cell lines, described in Chapter 2 as cells modified for unlimited growth *in vitro*. In about four years, in 1984, the group's initial reports of purification of killer proteins, of mediators endowed with the ability to execute cancer cells, began to appear. By culturing a mountain of lymphocyte- and monocyte-like cells and collecting hundreds of liters of the fluid in which those cells had been bathed, the researchers purified and characterized two proteins, calling lymphotoxin the one made by lymphocytes and TNF the one produced by monocytes. These two mediators were both rapidly cloned (almost simultaneously by many research teams) and shown to be generated also by real-life lymphocytes and monocytes isolated from people's blood, not just by artificially manipulated cell lines. "It's been cloned, therefore it exists," Lloyd Old wrote while reflecting on the identification of TNF, whose name he had originally coined.[11]

Immediately after cloning of TNF, Drs. Cerami and Beutler recognized the identity of this tumor-killer cytokine with their own cachectin, as mentioned above. Though it never completely disappeared, the term cachectin slowly faded from the literature, with TNF eventually taking hold as the cytokine's designation. However, neither of the two original names turned out to appropriately describe this body message. The cachectin line of investigation mostly developed into research aimed at understanding the role of TNF in inflammatory responses, whereas the cancer side took the direction of exploring the involvement of TNF in determining the fate of cells.

We can now follow the cachectin/inflammation path by reviewing studies that examined the involvement of TNF in the pathogenesis of one of the direst outcomes of an infection, septic shock.

Shock

Shock is a life-threatening condition that occurs when cells do not receive enough oxygen and nutrients because blood flow to various parts of the body becomes insufficient. This situation can damage multiple organs—kidneys, lungs, brain, liver—often irreparably. Shock can have many causes, such as inability of the heart to pump blood properly or major hemorrhages that sharply reduce the amount of blood in the circulatory system. A sudden and drastic reduction in blood pressure can also cause shock. The most familiar of these low-pressure forms is probably anaphylactic shock, which takes place in some people with allergies and can be fatal if not immediately treated.

Infections that run out of control are at the origin of another deadly form of shock, called septic shock. For still unclear reasons, a localized infection—usually of bacterial origin—can evolve into a generalized inflammatory response called sepsis, mentioned in the previous chapter when reviewing IL-1Ra. When even sepsis gets out of control, septic shock ensues. Millions of people all over the world die of septic shock every year; death rates are stunning even in the presence of the best health care. No specific therapy is available for sepsis and septic shock, with physicians to this day being limited to using all the technology they have (respirators, dialysis machines, intravenous fluids, antibiotics, etc.) to try to keep patients alive until the crisis is over.

The search for the causes of sepsis and septic shock has a long history. In time, investigators determined that administration of endotoxin to laboratory animals (the tool Dr. Cerami had used to identify cachectin) mimicked many of the characteristics of septic shock, including drastically reduced blood pressure, altered respiratory rate, weight loss, and death. Experimental induction of endotoxic shock thus became popular among scientists who wanted to model human septic shock in animals to try to understand the pathological mechanisms underlying this condition, with the ultimate goal of devising strategies to either prevent or treat it.

Drs. Beutler, Cerami, and their colleagues, as well as other investigators, had determined that macrophages exposed to endotoxin produced TNF, and knew that injection of endotoxin made TNF appear in the bloodstream. Dr. Cerami's team had also injected purified TNF into mice, observing weight loss and other effects that partly resembled endotoxic—and thus septic—shock. Now, armed with antibodies they had generated by immunizing rabbits with TNF (just as Charles Dinarello had done to make antibodies against IL-1), Cerami and colleagues took a subtraction approach to understand whether TNF was necessary for development of septic shock. They injected mice with their anti-TNF antibodies, thus preventively neutralizing this cytokine's activity, and then induced endotoxic shock by injecting mice with a bolus of endotoxin. In their highly influential 1985 publication in *Science*, Drs. Beutler and Cerami, together with Dr. Ian Milsark, reported that blocking TNF had an impressive protective effect: whereas only 50 percent of the mice receiving the endotoxin bolus survived, all the animals in which the scientists had neutralized TNF prior to injection of endotoxin lived.[12]

Subsequent studies demonstrated that this strategy also worked against live bacteria, not just bacterial fragments such as endotoxin. And it worked in baboons, which are definitely much more similar to humans than mice. Additional investigations supported these findings, as did multiple studies in humans showing that blood concentrations of TNF were elevated in patients with sepsis and septic shock. Many felt it was time to check whether blocking TNF would be an effective treatment in real patients with sepsis and septic shock.[13]

Indeed, in May 1990, the journal *The Lancet* published the results of a clinical trial that assessed the safety of neutralizing TNF in a small sample of fourteen patients with septic shock. No adverse reactions were noted; larger, properly designed studies looking at efficacy could therefore begin.[14]

There was much trepidation among those working in the field. I was then about to enter the world of TNF research myself, under the guidance of Pietro Ghezzi at the Mario Negri Institute in Milano. Not that I could appreciate any of the details at the time, my biology degree still fresh, but I vividly remember the discussions among senior investigators, the bubbling excitement that something historical was about to happen. This would be the first demonstration that directly intervening in the chain of inflammatory events by attacking a single cytokine could change the outcome of deadly septic shock.

In the following years, a few reports demonstrated positive—if transient—results on some outcomes, keeping alive hope that a treatment for septic shock would be at hand despite early warnings that things might not go as expected. Then, in 1995, the blow: a study that included almost a thousand patients at thirty-one hospitals in the United States and Canada showed no improvement in survival in those who had received antibodies against TNF compared to those who didn't. A couple of years later, a comparable trial demonstrated lack of efficacy of blocking IL-1 with IL-1Ra in patients with severe sepsis, as briefly mentioned in the previous chapter.[15]

Much debate over what might have gone wrong followed this deflating news. Timing of treatment was a big issue. Investigators, in fact, can easily pretreat experimental animals with whatever compound they are testing—such as anti-TNF antibodies or IL-1Ra—before applying the intervention aimed at mimicking human disease, an injection of endotoxin or a bolus of bacteria in this case. But this kind of pretreatment is obviously impossible in humans with septic shock, where the disease has already started by the time physicians can administer a medication. Most of the animal studies had shown beneficial effects when researchers had given anti-TNF before disease induction; the drug could still work if injected a couple of hours later, but efficacy started to wane. Many scientists soon began analyzing the clinical trial's disheartening results, noting that by the time patients are admitted to an Intensive Care Unit with sepsis or septic shock, the therapeutic window may already have closed, meaning that it may already be too late to intervene by blocking TNF (or IL-1). Researchers involved in preclinical studies have since become much more cautious about drawing conclusions from animal experimentation that uses pretreatment interventions, even though the issue of timing of drug administration in relation to disease induction in experimental models remains open.[16]

Whether investigators were using animal models that properly mimicked the specific human disease was another issue of contention. Sepsis and septic shock do not arise from the sudden appearance in the blood of a single bacterial product, or even a bolus of bacteria, but rather result from the improper response to a local infection, which may or may not spread to the rest of the body. Using a system in which a sepsis-like syndrome originated from a perforated intestine (a real cause of sepsis in humans), investigators in 1992 had shown lack of efficacy of TNF blockade, just as later happened in clinical trials. Could it be that researchers had used inappropriate experimental models to predict the desired intervention would be effective in humans? Controversy as to which of the available animal models—if any—properly reproduced the characteristics of sepsis and other types of critical illness has been raging ever since, and shows no signs of abating.[17]

Those who were trying to make sense of the trial's negative results, however, did not simply point the finger to inappropriate preclinical data. The ballet of mediators involved in the pathogenesis of sepsis and septic shock, in fact, turned out to be much more convoluted than anticipated. Some even questioned whether the mere idea of stopping the cascade of events by blocking a single cytokine might be wishful thinking. So many cytokines and other body messages are turned on in sepsis and septic shock, and so many of them have overlapping effects, that removing just one might not make much of a difference. There was also the observation that the body actually tries to control inflammation on its own during sepsis, with production of extremely elevated levels of cytokine antagonists like IL-1Ra and many others. So, did it make sense at all to push the amount of these already present cytokine antagonists even higher?[18]

Other issues were objects of discussion, such as selection of proper criteria for enrollment of patients in the clinical trials, the doses of therapeutic agents administered, and more. However, the inevitable fallout from these studies was that companies mostly retreated from a field—blocking pro-inflammatory cytokines to avoid death of patients with septic shock—that had become synonymous with failure, while scientists went back to the bench to try to figure things out. Interest in inhibition of TNF and other cytokines moved to a very different class of diseases, mostly chronic inflammatory conditions like rheumatoid arthritis. Despite these problems, the study of septic shock further consolidated awareness of the central role TNF plays in inflammation and helped to define the mechanisms of its action.[19]

Meanwhile, a separate set of investigators was concerned with the ability of TNF to kill cells rather than people.

What We Talk About When We Talk About Necrosis

Do the terms tumor and necrosis in TNF's name truly reflect this cytokine's main biological function? As always, before any question can be answered the exact meaning of the terms involved needs to be defined. This is particularly true in science, and even more in the field of cell death, where researchers are constantly rearranging concepts in the attempt of making sense of experimental evidence. While the term tumor in TNF's name does not require lengthy explanations, we instead need to elaborate in some detail on the exact meaning of necrosis.

Historically, clinicians and investigators used the word *necrosis* to refer to death of tissues, gangrene—with limbs, nose, ears or other parts of the body becoming black and obviously dead—being the clearest example. In the section on nutrition of his wonderful *Lectures on General Pathology*, Julius Cohnheim in 1898 defined necrosis as the "extinction of metabolic processes," citing loss of blood flow, extremes in temperature (either too high or too low), and inflammation secondary to infections as the underlying causes of necrosis. "When portions of tissue come into contact with decomposing urine, with putrid organic substances, or with offensive wound secretions, they usually die with more or less rapidity and to a greater or less extent. This very often happens indirectly as the result of inflammation going on to necrosis," Cohnheim wrote. "It is as though these substances were endowed with a capacity to inhibit the physiological metabolic processes. To which of the components of the fluids and materials the power of destroying the metabolism should be ascribed cannot be precisely stated at present." Cohnheim's description of necrosis in terms of metabolic disruption with a major involvement of inflammation contains fundamental elements of our contemporary understanding of cell death. However, his emphasis was firmly on tissues, since he completely discounted the importance of even attempting to examine individual cells. In fact, he wrote, "We have here to do only and solely with necrotic processes, and therefore nothing could well be more erroneous than to look upon the changes in the form and appearance of the tissue-cells as the expression of some vital processes or other."[20]

Later on, when the cell became the unit of investigation, scientists used the term necrosis to define accidental cell death as opposed to the programmed cell death of apoptosis briefly discussed in the previous chapter. Necrosis was messy death brought about by any sort of calamity, apoptosis a well-planned, orderly demise. Whereas apoptotic cells died a peaceful death that did not activate the immune system (actually, it might deactivate it), necrotic cells would spill out their content, perceived as danger by the immune system, with ensuing inflammation. This early-1970s classification of types of death was based on morphological characteristics, that is, on the typical microscopic changes that occur in cells as they undergo one type of death or the other. Still in 2005, the first official definition of the known variants of cellular demise, by the Nomenclature Committee on Cell Death, maintained morphology as the basis for defining types of cell death: necrosis continued to be considered an accident, not a programmed, orderly event.[21]

But molecular understanding of cell death was becoming more and more sophisticated and scientists began to question whether morphology was an appropriate way of classifying ways in which cells could die. Evidence that regulated molecular steps were not the exclusive domain of apoptosis but could also be part of the necrotic process slowly seeped in, changing the paradigm that saw necrosis as a mere accident. The neat lines distinguishing the ability of the two types of death to differentially affect the immune system also became fuzzy.[22]

In January 2015, the Nomenclature Committee on Cell Death generated a new set of recommendations. Accidental cell death should now only refer to death caused by a severe insult (physical trauma, harsh chemicals, extreme temperatures, etc.) that caused immediate death by disassembly of the cell. Every other type of death should be grouped under the new designation of regulated cell death, so-called because it requires participation of the cell's own molecular machinery. Implicitly, thus, the committee got rid of apoptosis, necrosis, and their numerous variations that had since popped up, recommending to shift focus from classification of types of cellular death to an understanding of the various phases a cell experiences on its way to death, and their molecular correlates.[23]

As a consequence of these modifications in scientists' interpretation of the term necrosis, the answer to the question "Does TNF cause tumor necrosis?" depends on the historical time in which the question itself is formulated. In 1975, when Lloyd Old and his colleagues had coined the term TNF, their criterion for defining necrosis was visual observation of subcutaneous tumors *in vivo* in mice. They also tested whether TNF could kill cells (both from tumors and not) *in vitro* by counting how many cells remained alive after exposure to samples that did or did not contain the necrotic factor.[24] Thus, the historically accurate answer here is yes, TNF does cause tumor necrosis if one uses the 1975 connotation of the term. In 1985, when Dr. Aggarwal and his colleagues cloned TNF, they defined its activity using what later became the standard bioassay for this cytokine, that is, killing of fibroblast-like cells under specific *in vitro* conditions. Here again TNF did kill cells, though not necessarily tumors (the investigators did not test that), so the answer to the same question may have already become less clear-cut.[25] Today, it is the question itself we need to reformulate to accommodate our current understanding of cell death, where necrosis as originally defined is no longer part of scientists' conceptual framework. Let's rephrase then, "Does TNF cause cell death?"

Tinef

To answer this newly formulated question, we turn to Dr. David Wallach, a professor in the Department of Biological Chemistry at the Weizmann Institute of Science in Rehovot, Israel, and one of the world's experts in the field. Indeed, Dr. Wallach's seminal contributions have been instrumental in shaping scientists' understanding of cell death.

During our Skype interview, him sitting in his office in Israel and me across the ocean, I first asked Dr. Wallach about his life and that of his family, before turning to the specifics of his research.

> I am part of the return of Jews to Israel. My family descends from Ezra the Scribe, the prophet that led the return of the Jews from their first exile in Babylon to Jerusalem in the fourth century BC; my parents returned to Israel again from Europe shortly before the Second World War. At that time, much of Israel was just dunes and rather wild, it was a developing country. Those members of my family that did not make it to Israel in time were murdered in the Holocaust. In Israel, in my childhood, one frequently encountered people with numbers from concentration camps tattooed on their arms, you could see people that came from the best universities in Europe, which I think contributed to the culture.
>
> I always liked biology. I got it from my parents, I guess, by genes and by education. I always had an attraction to science, I cannot tell you a specific event because it has always been there.
>
> I went to Hebrew University. My mentor for the PhD was Michael Schramm, who studied hormonally induced secretion of proteins and who, together with Zvi Selinger, introduced research on intracellular signaling to Israel and made important contributions to the field. Then I went to the NIH, in the United States, and studied cancer biology, the molecular mechanisms of cancer, with Ira Pastan.
>
> I came back to Israel in 1977, at the Weizmann Institute for Science, and I asked myself what would be the least known field where I could make the greatest contribution to science. At that time, cell

death was not studied. The mere idea that this could be a scientific question was not even phrased. I was trying to find something on which there was very limited knowledge, and I thought that cell death could be a worthy topic. So, I came to my field because I cared to contribute to an area that others did not study.[26]

When Dr. Wallach entered the field of cell death in the late 1970s, the distinction between necrosis and apoptosis was only a few years old and the importance of these processes in biology was not yet widely appreciated, to say the least.[27]

David Wallach pressed on anyway, and it was the search for experimental strategies to study cell death that brought him to TNF. As he told me during our interview and also wrote in an article chronicling his research on TNF, "People joked and called it *tinef*, which means dirt in Yiddish. All this looked very funny and so everybody studying TNF faced skepticism."[28]

Starting out by studying interferon-γ, a cytokine we already met, Wallach's team discovered a substance—a cytotoxin, as they called it—that could both kill cells and protect them from death. As you can imagine, these findings were quite puzzling, perhaps justifying some of Wallach's colleagues' initial skepticism. "Nobody knew what I was talking about," Wallach told me. "It was really very difficult. However, we developed an antibody against the cytotoxin and managed to apply it for isolation of the cytokine. This allowed us to confirm that the cell killing and the protection from it were both mediated by the same molecule." At this point, researchers at Genentech and elsewhere cloned TNF, as we reviewed a few pages back. Just as it had happened with Dr. Cerami's cachectin, David Wallach was able to recognize his cytotoxin in the new cytokine. He now had a defined molecule he could use to study mechanisms of cell death.

As we know, cytokines deliver their message to target cells by binding to specific receptors. David Wallach led one of the teams that identified the receptors for TNF. His group then proceeded to dissect the signaling pathways TNF activates when it binds to its receptors, paying particular attention to those that might be involved in cell death, which remained Dr. Wallach's central interest. In the mid-1990s, these studies led the investigators to an enzyme, which they named MACH, that looked suspiciously similar to what was then still known as IL-Iβ Converting Enzyme (see Chapter 3). By then, the study of cell death had become rather popular, and two other teams arrived at the same molecule at around the same time by using slightly different experimental approaches. The names they chose—FLICE and Mch5—were different from the term picked by Dr.Wallach, but the enzyme was the same. In just a few months, the caspase nomenclature was introduced, as we saw in the previous chapter, and MACH/FLICE/Mhc5 was renamed caspase-8, the molecule that sits at the very core of most cell death pathways.

Many additional significant findings soon followed identification of caspase-8, resulting in definition of a family of receptors, the TNF receptor superfamily, that is critically involved in determining the fate of cells. Does this mean we can finally answer our reformulated question and state that TNF causes cell death? We are getting closer, but a simple yes won't suffice.

Binding of TNF to its receptor activates a molecule called NF-κB in the target cell. This process directly triggers the inflammatory program, but also favors survival of that cell. Therefore, rather than death, as the cytokine's name and parts of its history would imply, the default response of a cell to TNF is the exact opposite, the cell will survive even in the face of signals that attempt to kill it. However, if a cell finds itself in a situation in which, for any reason, NF-κB cannot be activated (say, you have taken a putative drug that blocks NF-κB), then the meaning of the TNF message completely turns around: rather than being pushed to survive, that target cell is instead forced down the death pathway. If you recall, this pattern of responses corresponds pretty well to Dr. Wallach's puzzling early observations that his cytotoxin could make cells both live and die: the default response to TNF is survival, with death being the outcome only under very specific conditions.

To add an additional layer of complexity, in those situations in which TNF delivers a death message, a cell may interpret this message in multiple ways, proceeding toward its demise by following alternative paths that correspond to the traditional definition of apoptosis, necrosis, and their nuances. The details of what determines which type of death a cell will choose after sensing TNF are still being worked out, but modulation of energy metabolism seems to be involved, as Dr. Cohnheim had anticipated at the end of the nineteenth century.[29]

Let's now go back once again to our question about the relevance of TNF's name—tumor necrosis factor—to its actual function. At high doses and under specific conditions TNF can indeed kill cancer cells. This death-inducing activity has found limited clinical applications for treatment of metastatic melanoma (skin cancer) localized to the limbs. However, the field of cancer research has made an almost complete turnaround with regard to TNF. Investigators interested in cancer are now mostly studying this cytokine for its ability to promote cell survival by activation of NF-κB, which facilitates cancer growth rather than suppressing it. Thus, our current view of TNF in relation to cancer is just about at the antipodes of the function that occasioned this cytokine's identification and naming, as scientists now mostly consider TNF as a mediator that makes cancer thrive, not die.[30]

To summarize, leukocytes (mostly, but not only, macrophages) synthesize TNF when microbes and other danger signals activate these cells. Then, TNF binds to its receptors, which are present on a variety of cell types, and initiates an intricate cascade of intracellular signaling. The outcome of this signaling depends on the specific context in which the responding cell finds itself, in ways that we do not completely understand. That cell may be prompted to die (in different ways), but more often than not it perceives TNF as a survival message and thus decides to stay alive even in the presence of other stimuli that would push it toward its demise. Finally—and crucially—cells respond to TNF with production of additional body messages and other activities that promote inflammation. However, like most of the other pro-inflammatory cytokines, TNF also sets in motion a cascade of anti-inflammatory events that eventually curbs inflammation, in part by inhibiting further production of TNF itself. The next body message in our series, interleukin-10, is exactly such an anti-inflammatory cytokine, generated in response to pro-inflammatory stimuli.

Interleukin-10

Cytokine Synthesis Inhibitor Factor, the original name of IL-10, says it all: here is a body message that shuts down synthesis of other cytokines, a mediator whose main job is to control inflammatory fires.

Just like IL-1 and TNF, IL-10 is the founder of its own family, which includes nine related cytokines. Members of the IL-10 family are excellent housekeepers, each performing complementary tasks that collectively maintain the integrity of the epithelial tissues that line our body cavities, particularly the intestine. Only one member of the family though, our main character IL-10, is endowed with special fire-extinguishing powers.

Made by many types of leukocytes when exposed to inflammatory cytokines, IL-10 feeds back on the process that induced its own synthesis, in a negative loop that inhibits production of most pro-inflammatory mediators, including IL-1 and TNF. Our current understanding of IL-10 places this cytokine at the core of the anti-inflammatory response, but it was immunologists interested in lymphocyte-mediated immunity, rather than researchers specializing in inflammation, who discovered this body message.[31]

A Scientist on the Move

"IL-10 has a very interesting story. Let's start with children with immunodeficiency," began Dr. Maria Grazia Roncarolo, whom we briefly met in the previous chapter, when I asked her to tell me the story of the discovery of IL-10. Born in Torino, Dr. Roncarolo attended Medical School in the same Northern Italian city. "During my fellowship in pediatrics, I worked with immunodeficient children, who died without hope at the time because we did not have any cure for them," she told me.

The immunodeficiency Dr. Roncarolo mentions refers to a group of genetic disorders that prevents the immune system from responding to microbes in an appropriate manner. The most common type, Severe Combined Immuno Deficiency, or SCID, has become rather famous through the movie *Bubble Boy*, the story of a young man who had to live inside a sterile bubble so he wouldn't catch any germ that might kill him.

Lymphocytes are the defective cells in SCID and the other types of immunodeficiencies Dr. Roncarolo has been involved with. Because lymphocytes recognize specific microbes and entice the immune system to attack them, children with these kinds of immunodeficiency have an extremely high risk of succumbing to overwhelming infections.

Lymphocytes, like other leukocytes, are made in the bone marrow. Nowadays, many children with genetic immunodeficiencies can be cured with bone marrow transplants, so that the healthy transplanted cells can substitute for the child's own faulty lymphocytes. However, this type of treatment was not yet well established in the early 1980s, when Dr. Roncarolo had to travel with a young patient from Italy to a specialized center in Lyon, France, as the only chance for a procedure that may save that child's life. The procedure to be performed in Lyon consisted in obtaining blood stem cells from the liver of a nonviable fetus (the liver gives rise to blood cells during early fetal development, then the bone marrow takes over) and transplant them into the immunodeficient child, in the hope that the fetal liver cells would mature into healthy blood cells, just like today happens with a bone marrow transplant. The procedure was risky, and not always successful, but there weren't many other choices.

Whenever possible, a transplant of immune cells needs to come from a close relative of the patient, a person who is matched for a group of proteins present on the surface of the body's cells. This match ensures that the transplanted immune cells do not recognize as foreign the body of the new person in which they find themselves in. If there is no appropriate match, the transplanted cells often react to their foreign environment by attacking their new body with such a relentless force that stopping them is almost impossible. This frightening, debilitating, and potentially lethal situation is called graft-versus-host disease, and is still a major issue for patients who need a bone marrow transplant but do not have a matched relative who can donate cells.

"In 1983, I went to Lyon for the first time with one of our patients who needed a transplant of fetal liver cells that could be performed only there," Dr. Roncarolo continued. "A few years later, in 1986, I moved to Lyon as an assistant professor, doing my clinical work at the hospital and continuing my research at an outstanding immunology laboratory sponsored by the pharmaceutical company Schering-Plough. In Lyon I met my husband, who is Dutch and also an immunologist. Since the very beginning, my professional life has been tightly

interconnected with my personal life, they are almost inseparable,"
Dr. Roncarolo explained. "While in Lyon, I noticed that some of the
kids we were able to cure with fetal liver stem cells surprisingly did
not develop graft-versus-host disease despite a donor mismatch. I
thought we should study what went on with these children. Even
though they were the exception to the rule, if we could understand the
mechanism that protected them we might use that knowledge to help
the other kids. I was definitely naïve at that point, but I immediately
understood that it wasn't the absence of attacking cells that prevented
these few children from developing graft-versus-host disease. The dangerous cells were indeed present, but something kept them suppressed,
inactive. I tried various options to find what that something might be,
but nothing worked. Two entire lab books of negative results."

Meanwhile, scientists at DNAX, the California institute owned by
Schering-Plough (the same pharmaceutical company that sponsored
Dr. Roncarolo's research in Lyon), were busy identifying a new cyto-
kine from cultures of mouse T lymphocytes.

There are two main types of T lymphocytes, called helper and cytotoxic. Cytotoxic T lymphocytes kill cells that are infected with intracellular pathogens such as viruses, while T helper lymphocytes—those
we are concerned with here—don't kill anything but instead help the
rest of the immune system function in an efficient and effective
manner. Helper T lymphocytes act just like an orchestra conductor,
coordinating the response of other immune cells. Instead of using a
baton, helper T lymphocytes conduct by way of cytokines, which carry
their directions to any cell ready to listen via expression of those cytokines' receptors on its surface.

Helper T lymphocytes themselves come in a few different versions.
Researchers in the late 1980s, when the events I am describing were
taking place, knew of two types, simply named Th1 and Th2. Investigators Tim Mosmann and Robert Coffman, at DNAX, had proposed
this paradigm and terminology on the basis of many important observations.[32] The main feature that distinguishes Th1 from Th2 lymphocytes is the set of cytokines these cells produce once activated. Cytokines made by Th1 lymphocytes, mainly our old friend interferon-γ,
promote an immune response that is appropriate to fight intracellular bacteria and protozoa, bugs like the mycobacteria that cause

tuberculosis and the trypanosomes that inspired the discovery of cachectin/TNF. On the other hand, the cytokines released by Th2 cells are most helpful when a person is faced with a worm infestation.

To be effective, the Th1/Th2 response needs to be tightly coordinated. If you are infected with mycobacteria or trypanosomes, it is helpful to direct all of your immune system's might against those bugs, and not worry about the worms that may not be bothering you at the moment. Better not to waste resources in mounting a type of response that is not appropriate to your current foe. Indeed, one type of response shuts down its counterpart: cytokines made by Th1 lymphocytes inhibit the activity of Th2 cells, while cytokines produced by Th2 lymphocytes inhibit Th1 cells. This is the key point in the present context.[33]

By 1988, researchers had demonstrated that interferon-γ was the Th1-derived cytokine that suppressed Th2 cells. However, the identity of its doppelgänger, the Th2-derived cytokine that inhibited Th1 cells, was still unknown (biologists like symmetry, just like physicists). It was precisely this cytokine that DNAX researchers Kevin Moore, Paulo Vieira, and colleagues were busy trying to fish out from cultures of mouse Th2 lymphocyte-like cells in the late 1980s. They succeeded, and in 1990 reported identification and cloning of what had until then been referred to as Cytokine Synthesis Inhibitor Factor and was later renamed IL-10.[34]

Now that they had mouse IL-10, DNAX scientists were facing the issue of finding its human counterpart. To accomplish this feat they needed to culture human cells that would synthesize high levels of the new cytokine, but they were not having much luck in this regard.

"In 1988, Schering-Plough offered all of us who worked in human immunology in Lyon to move to DNAX in California," Dr. Roncarolo explained. "It was a very difficult decision for me, because DNAX was a research institute, without a hospital or clinic, and I was a practicing physician. After a long agony, I decided to jump, moved to Palo Alto with my husband, and ended up working at DNAX for almost nine years. When I arrived at DNAX, mouse IL-10 had just been discovered as a cytokine produced by Th2 lymphocytes that was able to block production of interferon-γ by Th1 cells. Paolo Vieira asked me if I had human cells that he could use to clone human IL-10. Without telling him the source, I gave him lymphocytes from one of the children from

Lyon that had not developed graft-versus-host disease. After only five days, he came to me and said he had cloned it, that the cells I gave him produced tons of IL-10."

The article reporting cloning of human IL-10 was published in 1991. But the cells Dr. Roncarolo had provided for cloning of human IL-10 were not Th2 lymphocytes, the source that had been used for identification of the mouse version of IL-10. They were instead a whole new type of lymphocyte.[35]

Regulatory Cells

In the 1970s, when immunologists where busy trying to understand the intricacies of lymphocytes' dynamics, investigators had proposed that some types of T lymphocytes might behave as suppressor cells that kept other components of the immune system under control, preventing them from causing trouble. However, the concept of suppressor lymphocytes remained vague, mostly because the experimental data in support of the actual existence of these cells were not truly convincing. The concept was therefore set aside for a while.

In the early 1980s, as one cytokine after another was being identified, the notion that T helper lymphocytes might come in different flavors began to emerge. This led to the definition of Th1 and Th2 lymphocytes that, as we just described, were instrumental in the discovery of mouse IL-10. Then, toward the end of the decade, the concept of populations of lymphocytes able to dampen an over exuberant immune system resurfaced, with the term *suppressor* replaced by *regulatory*.[36]

Together with Dr. Roncarolo, the scientist who most significantly contributed to the field of regulatory T lymphocytes is Dr. Fiona Powrie, who currently heads the Kennedy Institute for Rheumatology at the University of Oxford, in England. Dr. Powrie began studying lymphocytes' diversity while still a PhD student in Oxford. In 1990, she published a highly consequential article together with her PhD advisor, Dr. Don Mason, describing the strikingly opposite characteristics of two populations of T lymphocytes. When injected into rats that lacked their own lymphocytes, the first of these two populations induced a "severe wasting disease with inflammatory infiltrates in liver, lung, stomach, thyroid, and pancreas."[37] Rats died. In stark contrast, nothing happened to the rats when researchers injected them with the second type of lymphocytes. Crucially, Drs. Powrie and Mason demonstrated that even very small numbers of the second type of lymphocytes could completely reverse the wasting disease induced by their aggressive counterparts, saving rats that would otherwise have been killed by the severe inflammatory condition. Just as the cells Dr. Roncarolo was studying in transplanted SCID children, these two populations of lymphocytes did not correspond to the Th1 and Th2 versions that had been defined by DNAX researchers. Even though Drs. Powrie and Mason were unable at the time to describe a mechanism by which protective lymphocytes controlled the aggressive ones, their article nevertheless strongly contributed to lay the foundation for demonstration of the existence of a population of regulatory T helper lymphocytes. "It was such as striking life-or-death issue that it influenced the rest of my career," Dr. Powrie stated in a 2010 interview.[38]

After obtaining her PhD, Dr. Powrie took on a postdoctoral fellowship at DNAX at about the time IL-10 had been identified. It wasn't long before the team she worked with, led by Bob Coffman, was able to demonstrate that IL-10 was a central mechanism by which regulatory lymphocytes controlled immune-mediated pathology.

Meanwhile, Dr. Roncarolo and her colleagues, also at DNAX, continued to investigate the T lymphocytes from SCID patients that had allowed human IL-10 to be cloned. Their studies, too, independently demonstrated that IL-10 plays a fundamental role in the activity of T regulatory lymphocytes. Despite both being endowed with regulatory functions, the T lymphocytes studied by Drs. Powrie and Roncarolo turned out not to be identical, likely representing variants of regulatory T cells with somewhat different functions; additional variants have since been described by various investigators.[39]

This combined knowledge eventually consolidated into the notion that small populations of lymphocytes keep the immune system in check, curbing its damaging potential. There is now extensive evidence that removal of regulatory lymphocytes leads to development of autoimmune and chronic inflammatory diseases in mice, and that, in contrast, supplementing animals with regulatory cells reduces severity of experimental disease and promotes tolerance to transplants. Currently widely accepted, the concept that the immune system contains tiny numbers of cells that actively and powerfully tame their reactive counterparts, pacifying the immune system in large part by production of IL-10, was revolutionary at the time of its introduction.

The Second Wave

In parallel with work demonstrating that IL-10 was an essential mediator of the restraining power of regulatory lymphocytes, scientists investigated the connection between IL-10, infection, and inflammation, showing that not only regulatory lymphocytes but also various other types of leukocytes synthesize high amounts of IL-10 in response to inflammatory mediators. The data they generated indicate that IL-10 usually rises during a second wave of message production, a wave that follows and is induced by the initial output of pro-inflammatory factors that includes TNF and IL-1. This second wave is essential in curbing inflammation, but can at the same time be troublesome when the organism has to cope with pathogens.

By disproportionally dampening immunity and inflammation, excessive IL-10 can in fact render antimicrobial responses ineffective, therefore increasing susceptibility to infections. Elevated concentrations of IL-10, for instance, slash synthesis of interferon-γ, which is essential to keep in check the mycobacteria that cause tuberculosis, so that this chronic infection is more aggressive in the presence of abundant IL-10. This situation is nothing but the real-world reflection of the *in vitro* approach that had initially allowed identification of mouse IL-10 as the Th2-derived cytokine that inhibits Th1 responses (which are characterized by production of interferon-γ), reiterating how even the most artificial and overly simplified *in vitro* experimental system can be instrumental in allowing understanding of the complex *in vivo* dynamics of disease pathogenesis. Hence, if a Th1-type of response—namely interferon-γ—is needed to adequately confront an infection, restraining IL-10 may be advantageous.[40]

There is, however, a flip side to the medal. Because much of the damage caused by microbes is actually a consequence of the body's own response through production of pro-inflammatory mediators, having IL-10 around can be beneficial during infections, as this cytokine can help restrict potentially harmful reactions. A relevant example here is experimental pertussis (whooping cough), where IL-10 is necessary to limit tissue injury caused not by the bacteria themselves but by the body's own endogenous mediators, including TNF.[41]

These two contradictory but interlinked sides of IL-10 can coexist in the setting of the same infection, as demonstrated by the response to trypanosomes. Due to a florid innate immune response, these bugs have a hard time growing in mice in which IL-10 has been genetically removed, indicating that production of IL-10 during trypanosome infection reduces the ability of the immune system to handle microbial growth. However, even though absence of IL-10 allows the immune system to better control the growth of trypanosomes, this extra hand does not lead to a better overall outcome in experimental trypanosomiasis, quite the opposite. In fact, despite a lower protozoan burden, mice that are unable to synthesize IL-10 rapidly succumb to the disease due to uncontrolled inflammation occasioned by unchecked production of pro-inflammatory mediators, including TNF. When it's inflammation that kills, not directly the microbes themselves, the fire-extinguishing powers of IL-10 therefore can come handy.[42]

One of the ways in which IL-10 can be both beneficial and harmful is by suppressing production of mediators that attract leukocytes to the site of danger, mediators that are our next characters.

Chemokines

How can you tell whether some part of your body is inflamed? You can go by the five cardinal signs of inflammation: redness, swelling, heat, pain, and loss of function. A nail that punctures your foot will make it swell and get so red, hot, and painful that you can no longer walk, your foot losing its function. The same will happen to the lungs of somebody who catches the flu or the joints of individuals with arthritis.

There are many situations, however, when inflammation is much subtler, to the point one hardly knows it's there. People who succumb to heart attacks, for example, almost universally have had inflammation in the walls of their arteries for decades, but they wouldn't know unless specific tests are run. Many types of illnesses are associated with development of chronic inflammation that is not as evident as the type characterized by the five cardinal signs listed above. In these cases, only an expert eye looking at the affected tissues can tell whether something is going on. That eye generally belongs to a pathologist who examines a thin section of tissue on a slide under a microscope.

Forsaking the Neighborhood

A few pages back, we met Drs. Powrie and Mason who, in describing the severe wasting disease that could be subdued by regulatory T lymphocytes, mentioned that rats developed inflammatory infiltrates in many organs. What did they mean exactly by the term "inflammatory infiltrate"? They were referring to accumulation of leukocytes in tissues, outside of blood vessels. Leukocytes are also called white blood cells because they travel around the body in arteries, capillaries, and veins as part of the cellular components of blood (the other cellular components are red blood cells and platelets). Accumulation of white blood cells outside of blood vessels, that is, infiltration of leukocytes in between the cells that make up tissues, is a sure sign that an inflammatory response is ongoing.

Whenever danger and damage are perceived in a tissue, sentinel cells quickly respond by producing body messages that reach nearby tiny blood vessels. These messages, which include TNF and IL-1, talk to the cells that line those small veins and capillaries, and induce changes that make the inner lining of the vessel sticky by increasing expression of proteins called adhesion molecules. Sort of like a chemical Velcro, adhesion molecules capture leukocytes as they flow by in the bloodstream. This process makes leukocytes slow down and eventually stop, stuck on the adhesive Velcro that is present exclusively on the inside of those vessels that run through tissues where the sentinels had dispatched inflammatory mediators. At that point, leukocytes squeeze out of the blood vessels and find themselves into the tissue proper.

Here is how Julius Cohnheim described the process of leukocytes leaving a small vessel in his *Lectures of General Pathology*:

A pointed projection is seen on the external contour of the vessel wall; it pushes itself further outwards, increases in thickness, and the pointed projection is transformed into a colourless rounded hump; this grows longer and thicker, throws out fresh points, and gradually withdraws itself from the vessel wall, with which at last it is connected only by a long thin pedicle. Finally this also detaches itself, and now there lies outside the vessel . . . *a colourless blood-corpuscle.* While this is taking place at one spot, the same process has been carried on in other portions of the veins and capillaries. Quite a large number of white blood-cells have betaken themselves to the exterior of the vessels, and these are constantly followed by fresh ones. . . . But this is not all. The extravasated colourless corpuscles distribute themselves, in proportion as their numbers increase, over a large area, forsaking the neighbourhood of the vessels from which they were derived.[43]

Because every kind of leukocyte can cross blood vessels, a pathologist's skilled eye is able to observe the presence of neutrophils, monocytes, lymphocytes, or any of the other less abundant types. The dynamics of extravasation, of leukocytes crossing vessel walls, have been worked out in extreme detail since Dr. Cohnheim's writings. In the setting of acute inflammation, neutrophils arrive first, followed by monocytes and lymphocytes. In chronic inflammation, instead, neutrophils tend to be absent, and the infiltrate is mostly composed of macrophages (as mentioned, monocytes change name to macrophages once they cross blood vessels) and lymphocytes. Each wave of cell extravasation is tightly controlled and each cell type plays its own role in responding to the danger that elicited its arrival, clearing microbes and dead tissue but also potentially causing damage.

As Dr. Cohnheim described, once outside the vessels leukocytes move away—forsaking the neighborhood—directed toward the signals that called them in. To get there, extravasated leukocytes follow a chemical gradient made of body messages produced by sentinels and other cells in the damaged tissues. These are messages that instruct leukocytes to move closer to danger, to the zone that needs their attention, a zone that is marked by a higher concentration of chemical stimuli that are called chemokines from their ability to make cells undergo chemotaxis, that is, movement toward a chemical stimulus.

An essential component of the inflammatory response, chemokines can thus be thought of as messages that allow leukocytes to reach their workplace. Some chemokines, called homeostatic, are constantly synthesized: their job is to regulate the traffic of leukocytes in and out of immune and hematopoietic organs. Other chemokines—those we are dealing with here—are only produced at times of need and in places of danger.

Making the World a Little More Beautiful

Even though scientists had known about chemotaxis for quite a while, the first chemokines were discovered by paying attention to the presence of impurities during the process of identification of IL-1. In an interview in the winter of 2014, Dr. Joost (Joe) J. Oppenheim, head of the Cellular Immunology Group at the Center for Cancer Research of the National Cancer Institute in Frederick, Maryland, told me about the search for the first two chemokines after recounting the tale of his own life, a life that itself began in a dangerous place at a time of need:

> I was born August 11, 1934 in Venlo, in the Netherlands. My childhood was that of a much-loved Jewish prince, unfortunately interrupted by the Germans who invaded the Netherlands and started putting restrictions on Jewish life. We all had to wear Jewish stars and I couldn't go to the public school. Then we learned there was a roundup, with Jews being picked up to be sent to "work camps." We didn't exactly know what this was, but we didn't want to go. With help from our Dutch doctor, it was "decided" I had scarlet fever. I was put to bed, and when the Germans came to pick us up there was a big sign on the door saying that we were quarantined. As a result the Germans went away, saying they would come again later.
>
> My father then ended up in the hospital and was in bed next to Mr. Jackes Heuvelmans, a Dutch gentleman from the little town of Reusel on the Belgian border. Mr. Heuvelmans agreed to shelter my brother and myself while my parents went into hiding; his wife suggested we obtain false papers with new identities and that was arranged—I learned after the war—by the abbot of a monastery across the border in Belgium. My new name was Jan Blijdorp and my brother's Kees Blijdorp. I found out many years later that this was the name of the Rotterdam zoo. We were ostensibly orphans of the Rotterdam bombing raid, Protestant boys who had been adopted as foster children

by this big Catholic Heuvelmans family who ran the bakery and a Bed & Breakfast in the village of Reusel.

On December 12, 1942, my brother and I took off our Jewish stars—with great fear—and went on the bus with little suitcases down to Reusel to join the Heuvelmans family. They had six children already; it was a very religious Catholic family. This experience as a "hidden child" was extremely important and formative for me as a person. We spent twenty months under assumed names; we could not go to school because the monks who ran the school were afraid of being executed if they took us in. The only education I had was from a monk who volunteered to come once a week and gave me lessons in the Catholic catechism. When it became apparent the war would last more than just a few months, we settled down as being child number seven and eight in the family. Luckily, we had enough food, since Mr. Heuvelmans was the village baker. At the end of September 1944, Reusel—the last town held by the Germans in Southern Netherlands—was finally liberated.

So, my brother and I managed to survive the war. Unfortunately, my parents had been caught and sent to concentration camps. My mother ended up in Theresienstadt and my father in Auschwitz, where he disappeared along with the other one thousand five hundred people on his train. Luckily, my mother survived the war and returned to the Netherlands to find us.

We came to the United States at the end of 1946. I couldn't speak a word of English. I was put in the fifth grade in school even though I was twelve and the first day I was absolutely stricken because I could not understand a word of English, I couldn't imagine how I would ever learn anything. Since I felt I had to make a success of my life in the United States, I was awfully upset and hysterical. Everybody reassured me I would learn English and everything would be all right, but of course I thought these were total lies, though it turns out they were correct.

I have recounted these childhood experiences because they explain my motivation and my drive, my need for security and desire never to be involved in a war again. As a consequence, I studied very hard in school. I was lucky to qualify for the Bronx High School of Science, an excellent school in New York City, and met some of the most intelligent people in my life at that school. I then managed to get into Columbia College.

I wanted to study why people are crazy enough to go to war to kill each other, kill children, and blame Jews for killing Jesus; I tried very hard to understand this. I was told I had to become a psychiatrist to obtain answers to these questions, but this required that I first had to become a physician. So I went to Medical School at Columbia University. During my senior year there, I studied for half the year at the psychiatric institute, where all the Viennese psychiatrists trained by Freud, including his daughter, were working. I was incredibly impressed by their understanding of people with various mental illnesses. The only problem was that each one had a different understanding of what these conditions were, and all of them made up different stories and solutions. I decided I really couldn't spend seven years learning everybody's opinion about what's wrong with someone. Even though I wasn't a scientist by any means, it sort of offended my sensibility that psychiatrists at that time approached problems in such a subjective manner. I was at a loss, and instead became an internist.

I did an internship and residency at the University of Washington in Seattle, I wanted to get as far away from New York as possible, and then I applied to the NIH because I learned you could fulfill your military service there and not have to go to war in Vietnam, which I definitely wanted to avoid. Because I had done a psychiatric subinternship, my qualifications to perform research were nil in any substantive field. However, the chemotherapy service was desperately looking for clinical associates and I was recruited.

I worked with Drs. [Emil] Frei and [Emil] Freireich, who are written up in the book *The Emperor of All Maladies* as having succeeded in introducing multi-dose chemotherapy as a partially successful cancer treatment, particularly for leukemia. But at the time I was working there, the treatment was not successful and I became very discouraged. I therefore decided I would do research to see if I could find a better way to cure cancer. I became willy nilly an immunologist; I trained myself, which is probably a very awkward, inaccurate, and inefficient way of learning. On the other hand, it resulted in my being somewhat of a rebel at the time because I did not have formal training.

I then spent a year in England and learned the value of using animal models to understand the immune system. I returned to the NIH, first at the Dental Institute where they were very interested in inflammation of the gums and then back to the Cancer Institute where I have spent the rest of my scientific life.

Part of that scientific life was spent identifying the first two chemokines, and the rest doing research that, in addition to many consequential discoveries, planted "many ideas and seeds, enough for his many scientific children and grandchildren to harvest and sow again to make the world a little more beautiful each morning," as Dr. Grace Wong wrote on the Web site announcing a symposium in honor of Joe Oppenheim's eightieth birthday in August 2014.[44]

Attraction

Dr. Oppenheim's research on chemokines began as something else, as an attempt to understand the biological properties of IL-1. There wasn't yet any recombinant, pure IL-1 and the preparations used to investigate this cytokine's activity were therefore dirty, meaning they contained other molecules in addition to IL-1, molecules that were for the most part still unknown. In collaboration with Dr. Edward Leonard, who was also at the NIH, Dr. Oppenheim noted that dirty preparations of IL-1 made monocytes and neutrophils move through a porous filter, in what is called a chemotaxis bioassay. When the scientists put monocytes and neutrophils on one side of the filter and their dirty IL-1 on the other side, the cells moved through the pores, seemingly attracted to the side that contained IL-1. That bioassay for chemotaxis had been developed in 1962 by Dr. Stephen Boyden as a way to quantify the movement of cells toward a chemotactic gradient and thus assess the presence, potency and specificity of any putative chemotactic factor. The filter Dr. Boyden devised became known as a Boyden chamber and is still in use today as an effective and inexpensive bioassay to study chemotaxis.

Despite the long-standing observation that cells "forsake their neighborhood" after leaving blood vessels, the molecules that attracted leukocytes toward the site of danger were just beginning to be identified in the 1960s and 1970s. Investigators had demonstrated that some bacterial products and blood components had chemotactic activity for certain types of leukocytes, but the idea that sentinel cells—which are leukocytes themselves—would release chemoattractants that summoned leukocytes from the blood was not even mentioned in articles reviewing the topic in the early 1970s. As cytokines began to be identified and characterized, scientists considered the possibility that, in addition to their many other functions, these body messages might also possess chemotactic activity.

"My lab was joined by Dr. Kouji Matsushima, who had biochemical training in Japan," Dr. Oppenheim continued. "I asked him if he could purify IL-1 using the chemotaxis assay as a guide. He proceeded to do this by collaborating with Dr. Teizo Yoshimura in Ed Leonard's lab. Teizo would do the bioassays for chemotaxis and Kouji would do the biochemical assays for purification. What they found was that the cleaner the preparations of IL-1, the less chemotactic activity there was, till they got to the point where finally IL-1 was cloned and a bit of the pure cloned material was provided to us and it had no chemotactic effect whatsoever." This meant that it wasn't IL-1 itself but some of the impurities, that is, one or more of the unknown molecules present in the dirty preparations, that caused the chemotactic activity the NIH investigators had initially observed.

"Kouji and Teizo, as a side project, decided to purify this impurity. It took them about a year to find something that we decided to call monocyte-derived neutrophil chemotactic factor. Ed Leonard and myself were very happy with this," Dr. Oppenheim explained. The team published their initial findings in 1987, and in 1988 reported cloning of that neutrophil chemotactic factor. Meanwhile, as it was, three other groups in Europe (led by Dr. Marco Baggiolini at the University of Bern in Switzerland, Dr. Enno Christophers at the University of Kiel in Germany, and Drs. Alfons Billiau and Jo Van Damme at the University of Leuven in Belgium) found the same chemotactic factor using approaches similar to the ones employed by the NIH team. As usual, each group gave the new molecule a different name. Eventually, this neutrophil-attracting chemokine was called IL-8, the only chemokine to ever enter the ranks of interleukins (and a designation that continues to raise eyebrows, since all other chemokines are classified separately from interleukins).[45]

Drs. Oppenheim and Leonard, with their colleagues Drs. Matsushima and Yoshimura, had also observed the presence of a different factor in their dirty IL-1 preparations, something that attracted monocytes instead of neutrophils. They ended up purifying and cloning that protein and gene, as well, but now working as separate teams. "There was a bit of a disagreement in the two groups, so Ed and Teizo went after the monocyte chemotactic factor on their own, and so did Kouji and I. We got the international patent and they got the United States patent, we all discovered the same thing," Dr. Oppenheim explained, describing identification of the monocyte-attracting chemokine that was eventually named monocyte chemotactic protein-1, or MCP-1 (Oppenheim did not reveal what the disagreement was about.)

"Chance is very important in science. You can have a program and plan to do certain things, but you have to keep your mind open to possibilities and alternatives. The chemokine field, starting with IL-8 and MCP-1, was based on accidental discoveries, with the idea that there must be a contaminant. We just went ahead looking for that contaminant and found it," Dr. Oppenheim concluded. By deciding to purify an impurity, they had found something utterly novel and far-reaching.

Rapid Evolution

The chemokine family turned out to be much more numerous than initially thought, with about fifty members and more than twenty receptors. Starting out with a mind-boggling array of names that defied any logic, the chemokine nomenclature has since been highly rationalized, thanks to the presence of chemical signatures that allowed bundling of these molecules and their receptors into a few subfamilies. The two larger families of chemokines are called CCL and CXCL followed by a number (the L stands for ligand) based on the presence in the chemokine's sequence of two specific amino acids, two cysteines, either near each other (the CC group) or intercalated by a different amino acid (the CXC group). Thus, IL-8 is currently called CXCL8, while MCP-1 became CCL2. Naming of chemokine receptors uses a similar rational system, these receptors being called CCR or CXCR depending on the main type of chemokine they bind.

Rationality, though, seems to stop here. Chemokines in the CXC family generally attract neutrophils, while those in the CC class tend to attract monocytes and lymphocytes, but there is so much overlap between the two subfamilies that a clear distinction is not possible. Another level of complexity derives from the extreme promiscuity of this family's members. Chemokines share receptors and receptors share chemokines, the one-on-one correspondence between ligand and receptor typical of most body messages not applying at all in this family. Thus, for instance, IL-8/CXCL8 binds to both CXCR1 and CXCR2, but these two receptors can also be activated by at least eight additional chemokines. Even molecules that are not chemokines, including some made by microbes, can bind to chemokine receptors, while some chemokines can activate a receptor but block another one, alternatively behaving as agonists or antagonists. Finally, target cells typically have multiple types of chemokine receptors on their surface, while cells that produce chemokines generally make multiple versions of these body messages. You get the picture, it's a system that seems designed to defy comprehension.[46]

Because mobilization of leukocytes via chemokines is essential to combat infections, the wide diversity of the chemokine system—of both ligands and receptors—likely results from evolutionary pressure to increase the number of these molecules when early humans and other animals were faced with the threat of new pathogens.[47] Microbes, however, have also taken advantage of this molecular bounty that might have evolved to fight them. A variety of viruses, for instance, contain genes that encode for viral versions of chemokine receptors, these molecules shaping leukocytes' migration to the viruses' own advantage. Other examples include exploitation of the chemokine receptor DARC as a point of entry into red blood cells by the parasites that cause malaria, while the chemokine receptor CCR5 is used by the immunodeficiency virus HIV (the cause of AIDS) to get access inside immune cells. If microbes exploit the chemokine system, investigators are paying them back by developing therapeutic strategies aimed at just those interactions. For example, the observation that mutations in the CCR5 receptor confer resistance to HIV led to development of drugs that are used for treatment of HIV-infected individuals.[48]

The field of chemokine research itself has rapidly evolved. From an exclusive focus on leukocyte migration, experts in the area now recognize the chemokine system as an essential part of the body's interactions with microbes. But the range of chemokines' biological functions does not stop there. Chemokines can promote or block the spurt of new blood vessels (angiogenesis), make some types of cells proliferate, and even induce production of pro-inflammatory mediators, thus being involved in many facets of the inflammatory response. Despite its complexity, modulation of the chemokine system is being explored as a therapeutic target not only in infectious diseases but also in cancer, autoimmunity, disorders of the gut, and many other conditions. Drugs that block one or more chemokine receptors have been developed, even though many have not been as successful as initially hoped, probably due to the extreme intricacy and redundancy of the system.[49]

While new anti-inflammatory drugs, such as chemokine receptor antagonists, are being generated, investigation about the mechanism of action of some of the oldest anti-inflammatory medicines also opened up fresh avenues, in part through identification of annexin A1, the body message we are going to meet next.

Annexin A1

You have a headache or a fever, perhaps a sore throat. You reach for the medicine cabinet, pull out the aspirin bottle and gulp down a couple of pills with a sip of water. In half an hour or so your symptoms improve. Your child develops a strong allergic reaction, you rush to the emergency room, where the physician administers a shot of steroids. Things quickly get better, you and your child can start breathing again. Aspirin and steroids: Have you ever stopped to think how these potent drugs work their magic?

Both aspirin (and its relatives) and steroids were used clinically for their anti-inflammatory effects long before scientists figured out the mechanisms by which they work. The story leading to discovery of our next body message, annexin A1, is part of the quest to understand how these drugs exert their therapeutic action.

How Drugs Work

Roderick Flower was born in Southampton, England, in the immediate aftermath of the Second World War. He followed his Royal Air Force father around the planet, particularly to the Far East, where the family lived for some years. "This was a great experience, but it had a deleterious effect on my education," Dr. Flower wrote in response to my questions, "because when I returned to the UK I was years behind my cohort due to lack of proper teaching facilities on overseas Air Force bases at that time. I suppose astronomy was my first love; gazing up at the night sky in the tropics filled me with the sense of wonder that I still feel when contemplating the complexity of the natural world. I became interested in science mostly because of the inspirational effect of some teachers, and I was particularly interested in mathematics. Oddly, biology was never in the curriculum in those days (the 1950s), which is extraordinary considering the huge breakthrough that had been made in that decade by Crick and Watson and all the others!"[50]

After finishing school, for a while young Rod Flower worked in a bank, but didn't find the job congenial at all. While looking for a new position, he saw an ad in a newspaper for a laboratory technician to operate a computer in a pharmacology laboratory. "In those days—the 1960s—computers were very crude devices compared to today and they required twenty-four-hour attention," he explained.

> I applied for the job mainly because I was interested in numbers and computing, but also because several of my friends had similar jobs and it seemed to be a career of the future. The job was in the Department of Pharmacology at the Royal College of Surgeons in London. At my interview, I was told that the computing job had gone, but that they were looking for somebody else to act as a general laboratory technician. They asked whether I still wanted to go ahead with the interview. Since I had no other jobs in prospect at the time, I said I would be happy to do so. In retrospect, it was the best decision I ever made. I knew nothing about pharmacology—I remember going back home to look up the word in the dictionary—and, in fact, I hadn't even studied biology at school, so I knew nothing about the topic at all.

In that department, Flower worked under the immediate direction of Dr. John R. Vane, a British pharmacologist who later shared the Nobel Prize in Physiology or Medicine with Drs. Sune K. Bergström and Bengt I. Samuelsson for their discoveries concerning prostaglandins and related substances. To make a long and complex story short and simple, among many other key discoveries Dr. Vane had figured out the mechanism by which aspirin and similar drugs exerted their therapeutic action.

In fact, in 1971 Dr. Vane demonstrated that aspirin works by inhibiting the enzyme that makes prostaglandins, the molecules whose production is induced by inflammatory mediators and that, among their many other functions, directly cause fever, as discussed in the previous chapter. In his 1982 Nobel lecture—a true ode to the bioassay—Dr. Vane mentioned how, at the time of his initial findings, other scientists were unveiling the participation of prostaglandins in inflammation and fever, the two areas of research (the biochemical pharmacology of aspirin and the pathophysiology of fever and inflammation) supporting each other. Dr. Vane acknowledges Rod Flower in the document that accompanies the lecture, mentioning him as part of a remarkably talented and productive group of colleagues.[51]

But I am getting ahead of the story, so let's return to Dr. Flower's narrative.

> After a few weeks on the job I suddenly realized this was a profession I wanted to follow. Since I had no formal qualifications that would admit me to university, I studied in the evening and on a day-release course to rectify the situation. I left the department in 1969 and went to Sheffield University; I then returned to the department as John Vane's PhD student, arriving back at a propitious time. Actually, I had never really "left" because I used to work there during all my university vacations, so I was aware of the work that had been going on with aspirin-like drugs—in fact, I had done some work on the problem together with John during one of my summer vacations. However, my return coincided with the publication by John of his seminal paper demonstrating that aspirin inhibited prostaglandin production. Together with a clinical pharmacologist, I was the first to show that this happened in humans too.

Working side by side, Drs. Vane, Flower, and other colleagues made rapid advances in understanding the biology of prostaglandins, a project Dr. Flower still considers one of the most exciting he ever participated in. He learned something else, too.

> Aspirin-like drugs had been around since the end of the nineteenth century and, whilst everybody knew what their pharmacology was, nobody had a satisfactory explanation for their mechanism of action. At that time many drugs seemed to fit into this category—their clinical utility was discovered empirically. This was an issue I returned to several times during my career.

I also learned that using incredibly simple tools it was possible to dissect out the mechanism of action of a drug almost down to the molecular level. This taught me that it is possible to understand how drugs work using simple techniques together with cleverly designed experiments.

The field of anti-inflammatory drugs changed dramatically after the aspirin story was published and I was very much involved in that. It got me off to a good start in my career and also gave me an interest in other types of anti-inflammatory drugs. At the time, we had a selection of drugs, all of which were used in patients to treat rheumatoid arthritis and many other inflammatory conditions, but until John's breakthrough idea in 1971, no one had the slightest idea about how any of them worked.

In 1973, John Vane announced that he was leaving the laboratory at the Royal College of Surgeons to take up the post of research director at the Wellcome Foundation. This was an unusual pharmaceutical company founded by Henry Wellcome, a far-sighted man who set up his company in the early twentieth century by recruiting excellent scientists and giving them as much freedom as possible to study topics of their interest. It was enormously successful, and Wellcome developed a wide range of vaccines, medicines for infectious diseases, and other products. Dr. Vane decided to take some colleagues with him to set up his own research team at Wellcome, and invited Rod Flower to go along. "The lab was a great place to work. It was a very flat organizational structure. Everyone talked science all the time and socially we were very close and coherent. It was a great atmosphere and a terrific breeding ground for ideas," Dr. Flower explained.

Having made a start on understanding how aspirin functioned, Dr. Flower turned his attention to the steroids glucocorticoids, among the most potent anti-inflammatory drugs available—then and still today.

Glucocorticoids are hormones made by the adrenal glands, which sit on top of our two kidneys. These hormones are involved in the stress response and in an incredible number of other physiological functions, regulating both metabolism and inflammation. Hormones from the pituitary gland as well as inflammatory mediators induce production of glucocorticoids, which then act on leukocytes by downregulating these cells' activities and thus inhibiting inflammation. Glucocorticoids, in fact, suppress production of multiple inflammatory molecules, including IL1, TNF, and prostaglandins, while they upregulate synthesis of anti-inflammatory ones, like IL-1Ra and IL-10.

Despite the fact that adrenal glands produce the body's own glucocorticoids, pharmaceutical companies synthetically manufacture modified versions of these hormones that are widely used to treat a whole range of inflammatory, allergic, and autoimmune diseases. These are powerful drugs, often the only ones that can ameliorate dangerous and painful conditions, but they carry severe side effects that range from increased susceptibility to infections to metabolic alterations, bone loss, and many more. Obviously, figuring out how these powerful drugs work would be a decisive step forward in designing versions that are less toxic when used therapeutically.

It was already known in the 1970s that some effects of glucocorticoids are direct, not requiring any intermediary step by the responding cell, the glucocorticoid molecules binding to their own receptors and directly shutting down transcription of a series of genes. However, other functions seemed to need production of an in-between molecule made by the responding cell, a phenomenon that caught Dr. Flower's attention. His team showed that glucocorticoids inhibited production of prostaglandins in lungs that had been isolated from guinea pigs and maintained *ex vivo*, and that the drugs' inhibitory effect was prevented if the investigators blocked protein production in the lungs. This meant that a new protein was needed for glucocorticoids to suppress prostaglandin production. "When trying to imagine how the glucocorticoids actually blocked prostaglandin release from the lung, I had in mind the fact that, since protein synthesis was involved, it was likely that some new protein was required for the effect. Of course, this could have been an enzyme or a signaling factor, but I was disposed to think it might be a soluble mediator and very much taken up with the exciting implications of that idea," Dr. Flower explained. This is a thought process analogous to the one that led to understanding that bacteria do not induce fever directly but need the mediation of cytokines (which then induce prostaglandins). However, in this case the inducer was not a foreign invader but a hormone made by the body (or a synthetic version made in a pharmaceutical laboratory) and the detection system was not a whole animal, like Charles Dinarello's rabbits, but an isolated lung.

> I had always been interested in the history of pharmacology (at the time I was contemplating writing a book on the subject, although it never really got off the ground) and in particular in the use of clever experimental techniques. One that always stuck in my mind was the experiment by Otto Loewi showing that acetylcholine was the agent responsible for slowing the heart. In this experiment, Loewi used two hearts in series and when he stimulated the vagus nerve of one heart, the second heart also slowed down, showing that something was released from the first heart that influenced the second.[52]
>
> The idea came to me when I was lying in the bath one night—I know it's a cliché but I think it's true that we have our best ideas when we are relaxed. Influenced by Loewi's acetylcholine experiments, I set up a system whereby I put two lungs in series. The second lung in the series was used as a detection system. My hypothesis was that if steroids were infused into the first lung in the series, they would cause the synthesis and release of the hypothetical protein, which would flow through the system and affect the second lung. But how could we prevent steroids from producing production of the hypothetical mediator in the second lung? The answer was to infuse inhibitors of RNA and protein synthesis into the second lung, so that the steroids could not produce their effect in that organ.
>
> Once set up, the experiment was quite easy, one simply infused steroids either directly into the second lung, in which case they had no effect on the release of prostaglandins [because protein synthesis had been blocked in that second lung], or you switched the infusion line so that the steroids went through the first lung, and we found to our delight that this did in fact produce a decrease in prostaglandin release in the second lung, exactly as predicted. This formed the basis of our 1979 *Nature* paper.[53]
>
> A few other researchers were obviously thinking along similar lines to us. First of all, the group of Françoise Russo-Marie in Paris. She had noted an effect of glucocorticoids on cultured cells and hypothesized that this might be due to the synthesis of a new factor. Massimo Di Rosa's group in Naples had a similar idea and used macrophages as a test system. The third was Julius Axelrod's group at the NIH. They had a similar, although not identical, idea based upon their work on neutrophils, showing that steroids could block migration.

We all became friends and published a joint paper suggesting that the factor we were studying was almost certainly identical. We also published several more joint papers. Actually, we became lifetime friends with these investigators and we still see a lot of the scientists involved in that discovery today. On the whole it was a very pleasant experience.

That 1984 joint paper introduced the term *lipocortin* for the new protein to consolidate the disparate nomenclature the various groups had used until then.[54] However, at this point Dr. Flower and colleagues only knew their mediator was a protein, perhaps even more than one, but had trouble purifying it (as many other researchers did back then). "In fact, I found myself at a bit of a dead end here," Flower explained.

A year or so later, he left the Wellcome Foundation attracted by the offer of a chair at the University of Bath and entered into a formal collaboration with scientists at the biotech company Biogen, in Boston, Massachusetts, for identification of lipocortin. "The people at Biogen were fantastic protein chemists and it wasn't long before they had cloned the gene, sequenced the protein, and this was really the start of the story, because without them it would still be an unknown soluble factor!" Dr. Flower asserted.

The article reporting cloning of lipocortin, by Dr. Barbara Wallner and her colleagues at Biogen, was published in *Nature* in 1986. Rod Flower was not one of the authors, but he wrote an accompanying editorial underlying the importance of the findings.[55] When the sequence of lipocortin was finally deciphered and published, several groups noticed that it had motifs in common with other proteins, leading to the idea that there was a family of proteins of which lipocortin was one member. The family was renamed the annexins. "Actually I didn't like the name very much at the time and wrote to *Nature* suggesting it was too soon to come up with a definitive nomenclature, but the name stuck. So, lipocortin became Annexin A1 (Anxa1)," Dr. Flower noted.

Whatever its name, here was a new body message—sequenced and cloned—that explained at least part of the activity of some of the most potent anti-inflammatory drugs available. One would think scientists involved in studying other inflammation-related mediators would take notice. However, whereas Anxa1 investigators seamlessly enmeshed their research with the field of cytokines and chemokines, outside of a few encounters that helped extend the scope of Anxa1's anti-inflammatory activities beyond its ability to inhibit production of prostaglandins, the world of cytokine research did not seem that much interested in studying Anxa1. The report of a 1989 conference in Siena, Italy, perfectly illustrates this parochialism. Using the metaphor of *Il Palio*, the medieval tradition of a horse race in Siena's central square that pits different city neighborhoods against each other, Drs. Beutler and Flower described how Siena's competing *contrade* could serve as a perfect metaphor for the scientific factions involved in research focusing on interleukins, TNF, and lipocortins, the three main topics of the conference.[56]

Ignoring sectarian issues, Flower continued to pursue his research on Anxa1. Soon he moved again, to what would become his beloved long-term scientific home.

> When I was at Bath, I found myself in a bit of a dilemma. Whilst I enjoyed the job, cutbacks in academic staff meant that my professorial colleagues and I were sharing a much higher burden of administrative activity than we should have been. This meant that my time in the lab was reduced (common story) and it got to the point that I realized that unless I did something about it, my research career would probably be finished.
>
> As it happens, John Vane had moved from Wellcome when the company was taken over by Glaxo in 1986 and had been offered an office in Charterhouse Square by the dean of [what was then] Bart's Hospital Medical School. Vane came up with the idea of founding a new institute—which became the William Harvey Research Institute—and wrote to me asking if I would like to join in in doing so. I was enthusiastic about this idea, as it promised me a way out of the dilemma I had found myself in. When I first got there, all I had was an office, but I quickly attracted some staff and we set up a small lab with four or five people.

Dr. Flower eventually rose to the rank of director of the William Harvey Research Institute. After serving his four-year fixed term, he left the reins to his successors and went on with his research. The institute, which over the years has grown to a staff of hundreds and is part of Queen Mary University of London, is currently codirected by

Dr. Mark Caulfield, a renowned cardiovascular pharmacologist, and Dr. Mauro Perretti, one of Flower's first hires at the institute. "We now know, from Mauro's work especially, that Anxa1 has extraordinarily potent anti-inflammatory activity," Dr. Flower concluded.[57]

How Molecules that Make Drugs Work Work

Mauro Perretti was born in Parma, in Northern Italy, but he likes to clarify that he is a southerner, as he lived in Manduria, a town in the south of the country, until he was nineteen. Like Dr. Flower, Perretti wasn't much interested in science as a child, preferring sports instead, and did not have clear career plans. "When I finished high school, I didn't know what to do. My cousin was going to Florence to study pharmacy. I said 'I'll come as well,' and went with her, ending up with a degree in medicinal chemistry," Dr. Perretti explained. "In Florence, I remember in particular Professor Franca Buffoni, who gave fantastic lectures in pharmacology; she was very inspirational," Dr. Perretti continued. After obtaining his degree, Perretti took an administrative job, but was not satisfied, just like Rod Flower had been unhappy in his first bank job. As soon as he had the opportunity, he quit his position and joined the Sclavo Research Center in Siena.

Founded in 1904 by infectious diseases specialist Achille Sclavo as a public institute with a focus on vaccines, in the 1980s the Sclavo center had developed into a research powerhouse in pharmacology and immunology.[58] "At Sclavo, I met Luca Parente, who was the head of the pharmacology laboratory, and I started to work on inflammation. I learned a lot from Luca. Actually, working at Sclavo was quite illuminating for me. I was rather provincial, while the lab directors at Sclavo were mostly Italians who had come back after long spells abroad. English was spoken quite a lot, but I was narrow-minded and complained about this," Dr. Perretti recalls. Originally from Naples, Luca Parente had worked with Dr. Di Rosa (who had independently discovered Anxa1, as mentioned above) at the university of that city and had directly collaborated with Rod Flower on the early characterization of Anxa1.

After getting a master's degree in pharmacology and toxicology, Perretti eventually changed his mind about the English language and decided he could not be a biomedical researcher without learning it. At the suggestion of Dr. Parente, he decided to go to Bath, where Rod Flower was based at the time, so he could take English classes in the morning and join Dr. Flower's laboratory in the afternoon. "So, I learned English, got to know Rod, and went back to Sclavo," Dr. Perretti continued. "If we fast-forward two or three years, my wife—who is also a pharmacologist—had to do an experience abroad as part of her doctorate and decided to do it in London."[59] For a second time, Perretti said "I'll come as well," moved to London with his wife, and joined Dr. Flower's team at the new William Harvey Research Institute. There, he focused on Anxa1.

"Already when I was at Sclavo, I had started working on models of cell migration, which was being dissected thanks to the discovery of specific adhesion molecules and chemokines," Dr. Perretti explained. He decided to study the involvement of Anxa1 in chemotaxis using a system called the air pouch model. This is a simple experimental setup, in which air is injected under the skin of mice or rats with an empty syringe, creating an air-filled pouch. After a few days, the researcher can inject substances into the air pouch and evaluate accumulation of leukocytes, sort of like an *in vivo* Boyden chamber (the system Dr. Oppenheim's team had used to identify IL-8 and MCP-1).

When a scientist performs chemotaxis research *in vivo*, rather than *in vitro*, the whole body of the experimental animal participates in the response, not just the cells and mediators selected for investigation in a Boyden chamber. This can be a disadvantage if one wants to study a single substance or cell type, but is advantageous if the goal is to examine the interactions of the various components of the inflammatory response. On the other hand, the air pouch model is a highly simplified *in vivo* system, as it does not attempt to mimic a real human disease with all its complexities. "I thought it was better to use a very neat, simple system, because there you could dissect the mechanism in a better way," Dr. Perretti explained. As always, the choice of the best experimental model depends on the exact question the investigator is asking: sometimes an apparently abstruse system, like the one that led to identification of IL-18, is the most appropriate, at other times an extremely simplified one is best, as in this case.

The air pouch model allowed Drs. Perretti and Flower to study the powerful effects of Anxa1 in modulation of neutrophil chemotaxis. As the investigators knew from previous studies, injection of IL-1 into the pouch attracted neutrophils (IL-1 induces synthesis of chemokines, which then induce chemotaxis of neutrophils), while administration of glucocorticoids inhibited this response. Now that they had their system set up, the researchers could attempt to answer their new questions, beginning by adding Anxa1 to the system. Thus, they injected mice with recombinant Anxa1, and showed that IL-1 was no longer able to induce chemotaxis, just as had happened when mice had received glucocorticoids. This addition approach clearly demonstrated that Anxa1 potently reduced chemotaxis, but was Anxa1 necessary for the inhibitory action of glucocorticoids? To answer this question, as we know, the researchers needed to use a subtraction approach. So, the investigators injected IL-1 and glucocorticoids together into the air pouch, but only after they had blocked the activity of any Anxa1 the animals themselves could make. Remarkably, glucocorticoids no longer suppressed chemotaxis in mice in which Anxa1 had been blocked, demonstrating that this body message was necessary for the drugs to exert their antichemotactic activity.[60]

The data Mauro Perretti collected during this and subsequent studies proposed that Anxa1 acted by directly affecting neutrophils rather than indirectly through inhibition of prostaglandins. Moreover, other studies indicated that neutrophils themselves produced and released large amounts of Anxa1, which fed back to modulate these cells' ability to cross blood vessels, thus acting as a true body message. "That was very heretical at the time, as often is the case. I have subsequently been told by an expert that they didn't believe that paper and they thought I was doing something wrong, either I was cheating or there was something wrong," Dr. Perretti told me, referring to his 1996 *Nature Medicine* article, the first report of an endogenous negative feedback process regulating neutrophil extravasation through an autocrine mechanism.[61]

Dr. Perretti's research on Anxa1 proceeded by investigation of the mechanisms by which this protein exerted its potent anti-inflammatory effects, the list of which grew longer by the day. If Anxa1 was a true body message, as Perretti's neutrophils data indicated, it must bind to and activate a receptor on target cells. While Perretti and his team were on the hunt for such a receptor, it was Drs. Antje Walther, Kristina Riehemann, and Volker Gerke, from the Institute for Medical Biochemistry in Münster, Germany, who ended up finding it. "At the beginning, I was a bit annoyed because clearly it was a scoop. On the other hand it was a fantastic opportunity for other people to carry on and jump on the wagon and start working on it," Dr. Perretti told me. The Perretti team soon confirmed and expanded these findings, their reports suggesting that other receptors might also be involved.[62]

Interestingly, the Anxa1 receptor was not a new protein but instead belonged to an already famous family, known as formyl peptide receptors. Then mostly studied for their ability to induce chemotaxis after recognition of microbial components, researchers later established that a diverse host of molecules can attach to members of this family of receptors. Indeed, shortly after demonstration that Anxa1 bound to a member of the formyl peptide receptor family, evidence emerged that other anti-inflammatory molecules, called lipoxins, attached to the same receptor. Lipoxins, however, are not proteins, like Anxa1, but lipids that resemble prostaglandins, which means that the same formyl peptide receptor can bind both a protein and a lipid, two radically different types of molecules. This is a rather unusual property, again something that initially raised many eyebrows.

Despite Anxa1 and lipoxins being structurally dissimilar, they are both part of the network of mediators involved in the resolution of inflammation, a molecular program that actively promotes return to homeostasis after an inflammatory insult. Whereas anti-inflammatory cytokines like IL-10 control inflammation mostly by shutting down pro-inflammatory pathways, pro-resolution factors instead act in large part by turning on the healing process. As part of the resolution response, the body naturally forms both Anxa1 and lipoxins, but drugs we discussed in this section can boost these molecules' production, aspirin increasing lipoxins, glucocorticoids Anxa1.[63]

The *Nature Medicine* article showing that pro-resolving Anxa1 and lipoxins share the same receptor was coauthored by Drs. Mauro Perretti and Charles Serhan, of Harvard University.[64] The two scientists had first met while in Florence for a conference. "I was coming back to the hotel around midnight from a dinner with friends from my university days," Dr. Perretti told me. "We met in Piazza della Stazione, I think Charlie was jet-lagged. We started talking and we carried on for three, four hours, until early morning. It was really mind-blowing." The two scientists reasoned that perhaps the pro-resolving molecules they were interested in, Anxa1 for Mauro Perretti, lipoxins for Charles Serhan, could converge at some point. Out of that nighttime conversation, two years later the *Nature Medicine* article was born, nicely bringing together the two medications that had started the whole story, from John Vane's findings on the mechanism of aspirin's therapeutic actions, mediated at least in part by lipoxins, through Dr. Flower's identification of Anxa1 by studying glucocorticoids in isolated guinea pigs' lungs.

Intestinal Balance

Now that we have all our characters cast, let us bring them together in the intestine.

Here is TNF, mighty dangerous with its inflammatory power but at same time needed to control the fate of cells, discovered by way of wasted African cattle and tumor-killing concoctions. Over there is IL-10, an anti-inflammatory cytokine that helps make sure the nice bacteria that share our intestines do not get misinterpreted for nasty invaders. It was analysis of the symmetrical chatting between lymphocytes as well as care of immunodeficient children that led to discovery of this fire-extinguishing cytokine. We've also got chemokines, ready to summon help in the form of leukocytes moving in from the bloodstream into tissues to fight pathogens, but sometimes causing unwarranted trouble. The first members of this family of cell-attracting mediators were identified by digging into the dirt of the initial, impure preparations of IL-1. And, finally, we have problem-solving Anxa1, which coordinates damage repair and was discovered by scientists trying to understand how anti-inflammatory drugs work.

A Delicate but Solid Equilibrium

Our bowel, particularly the colon (the last part of the intestine where stools form), thrives under a delicate—yet remarkably solid—equilibrium. The intestine is home to the microbiota, billions of microbes that are a necessary component of our organism, contributing in innumerable ways to our health and well-being but at times turning dangerous.

A single layer of epithelial cells, just a fraction of a millimeter thin, separates the content of the bowel, with its mixture of bacteria and decomposing organic matter, from our true inside, our tissues proper. Structures made of specialized proteins keep these epithelial cells tightly attached to each other, controlling the flux of material from the intestinal lumen to the underlying tissues and vice versa. This thin layer of cells is constantly being renewed, with old cells replaced by new ones every few days following a highly organized pattern.

The intestine is also home to the most extensive network of immune cells in the whole body. Looming just underneath that single layer of epithelial cells, here and there intercalating with it, are countless leukocytes of many different types, which communicate with the microbiota and the epithelial cells in a busy exchange of information whose overall outcome is an ever-emerging equilibrium that constantly reestablishes itself.

The coordinated dance of body messages—including the ones we have just reviewed—is crucial in maintaining this intestinal balance, ensuring proper cell renewal as well as appropriate lines of communication between our own cells and the microbiota. As fickle as such a system might seem, its inherent flexibility makes it dependable in the vast majority of circumstances. Not always, though. As is often the case, examining situations in which balance is lost can help us understand the components that are essential to keep things working smoothly.

When Balance Is Lost

Inflammatory bowel disease (IBD) is a condition characterized by presence of chronic intestinal inflammation, most commonly located in the lower part of the gut. Individuals affected by IBD have abdominal pain, diarrhea, and a host of additional issues that may come and go, relapsing and remitting but usually not resolving on their own.

Extensive evidence points to disruption of the cross talk between microbiota and immune cells in the bowel of genetically susceptible persons as the most likely cause of IBD. In truth, just like with rheumatoid arthritis, nobody really knows what causes this chronic illness. Although we do not understand the ultimate cause of IBD, research on body messages has led to development of therapies that effectively control disease in a large percentage of patients, again in parallel with rheumatoid arthritis.

As promised at the beginning of the chapter, it is now time to proceed to appreciate the ways in which the body messages we introduced in these pages fit into the picture of IBD and, more generally, of intestinal balance.

In the mid-1980s, investigators demonstrated that the intestinal wall of patients with IBD contains increased numbers of leukocytes and elevated amounts of prostaglandins compared to the bowel of healthy persons. In the early 1990s, additional studies, mostly originating from the group of Dr. Thomas T. MacDonald at St. Bartholomew's Hospital in London, also pointed to the presence of elevated levels of TNF in the gut of IBD patients. This knowledge, supplemented by additional evidence, rapidly developed into clinical trials, showing that an antibody that blocked TNF dramatically reduced disease severity in individuals affected by IBD. The first small study—in just a handful of patients—took place in the mid-1990s at the Amsterdam Academic Medical Center in the Netherlands, at about the same time anti-TNF treatment was failing in the septic shock trials we reviewed a few pages back. Originally developed in the laboratory of Dr. Jan Vilcek at New York University, the anti-TNF antibody used in this first small IBD trial came from what was then Centocor, a Pennsylvania biotech. In 1998, that antibody became the drug Infliximab, the first in a series of TNF-blocking agents that revolutionized treatment of IBD, rheumatoid arthritis and other diseases.[65]

However, not every patient responds to anti-TNF therapy, and some who initially get better later stop being responsive. Clearly, blocking TNF is sufficient in many cases of IBD but not always. One should remember here that TNF is not only a pro-inflammatory cytokine but is also closely involved in modulation of cell death and survival. In the intestine, in particular, appropriate levels of TNF are important in maintaining the organized program of epithelial cell turnover mentioned above. Moreover, although high quantities of TNF are harmful, this mediator is required for effective repair of damage to epithelial cells. Finally, because TNF is crucial in shaping the immune response against pathogens, susceptibility to infections increases in patients who take drugs that block TNF.[66]

It is therefore no surprise that additional and alternative treatments for IBD other than blocking TNF have been developed, while new ones are being sought. Some pharmaceutical companies are attempting to block a few of the other pro-inflammatory mediators that are elevated in the intestine of IBD patients, while others are looking downstream. Upregulation of chemokine synthesis is one of the mechanisms TNF and other cytokines share in causing inflammation, chemokines then attracting leukocytes from the blood to the gut. There is indeed plenty of evidence demonstrating presence of elevated levels of many chemokines in the bowel of IBD patients; more than one company is busy assessing whether inhibiting chemokines' action by blocking their receptors might be a valuable therapeutic option. However, because chemokines are vital in the fight against microbial pathogens, this type of treatment is likely to carry the same side effects as anti-TNF drugs, namely, an increased susceptibility to infections.[67]

Rather than going downstream, one can instead move in the other direction and ask what leads to upregulation of TNF and the other pro-inflammatory mediators in IBD. Despite the ultimate cause being unknown, disruption of anti-inflammatory pathways may be an important contributor to the gut inflammation of IBD, with IL-10 possibly playing a central role.

Already in 1993, scientists demonstrated that mice engineered to lack the IL-10 gene spontaneously developed a type of chronic intestinal inflammation that resembled human IBD. In fact, the bowel of these mice was infiltrated with numerous leukocytes, indicating upregulation of chemokine production, and plenty of TNF could be found. Remarkably, when investigators raised these IL-10-deficient mice under conditions in which the microbiota was completely absent, the animals developed very little intestinal inflammation, if any. This observation clearly indicated that otherwise harmless—even beneficial—microbes were able to get the upper hand and cause inflammatory trouble by upregulating production of TNF and other body messages when IL-10 was not there to sing its calming lullaby.

Importantly, the bowel-protective activities of IL-10 are not limited to mice. Genetic studies have shown that mutations in the IL-10 receptor cause a severe form of IBD that develops in babies who are just a few months old, a period when the gut immune system begins to have a meaningful conversation with the microbiota. Though extremely rare, this is one of the most dreadful types of IBD, resistant to all available pharmacological therapies. These extremely sick infants can only be cured by bone marrow transplantation, so that immune cells with a functional IL-10 receptor can substitute for the babies' own faulty cells (this is, by the way, a type of intervention similar to the one used in children with SCID that had been at the origin of the discovery of human IL-10).[68]

Spontaneous, chronic, potentially lethal bowel inflammation in babies and mice who do not have a functional IL-10 system: a stronger proof of the absolute necessity of this cytokine in maintenance of intestinal homeostasis could not be asked for. And yet, despite these remarkable observations, whether alterations in IL-10 production or activity are actually involved in the pathogenesis of the most common forms of IBD remains under intense debate. Investigators have attempted—and failed—to treat IBD with administration of IL-10. Although the reasons for this failure may have nothing to do with the underlying biology but rather be related to the way trials have been implemented, excessive levels of IL-10, as we previously discussed, increase susceptibility to certain types of infections. Therefore, even if successful in dampening intestinal inflammation, supplementation with IL-10 may possibly result in the same troubles seen with anti-TNF drugs: more risk of infections.[69]

Perhaps, interventions that target our fourth body message, Anxa1, may skirt these issues. Since Anxa1 is mostly involved in the active resolution of inflammation rather than in suppressing immune responses, boosting its levels may promote healing without increasing risk of infections. In the early 1990s, evidence that Anxa1 might be involved in IBD began to emerge from work of Dr. Youichirou Sakanoue and colleagues at Hyogo College of Medicine in Japan. Since then, various investigators demonstrated involvement of Anxa1 in IBD, but the field overall remained sluggish compared to the rapid accumulation of data on the role of TNF and other cytokines, which allowed rapid translation into pharmacological treatments. Recent work by Dr. Asma Nusrat at Emory University, in collaboration with Mauro Perretti, as well as parallel studies by Dr. Basil Rigas at Stony Brook University indicate that administration of Anxa1 accelerates intestinal healing in mice. However promising, this evidence is still very preliminary; whether Anxa1 will work in real patients, and do so without generating dangerous side effects, remains to be seen.[70]

There Are No Bad Kids . . .

This chapter's overview of a small subset of body messages allowed us to highlight the unique history of the discovery of each molecule and of the investigators involved in those discoveries. While doing that, these pages also delineated the specific role each mediator plays in our organism. The evidence reviewed indicates how, through their intertwined actions, both pro- and anti-inflammatory mediators are necessary to maintain that flexible and constantly renewed balance we call health. Absence or overabundance of any of these mediators, as well as presence in the wrong place or at the wrong time, can cause trouble. Just like the Italian kids of the TV ad of my childhood, there are no bad body messages, only undisciplined ones.

5

Traps and Diversions:
Cytokine Decoys

THE WORD DECOY awakens images of a wooden duck floating in a pond, a watchful hunter waiting nearby, rifle in hand, ready to take down feather-and-flesh birds attracted by their phony sister. The term, however, originally indicated not a fake duck but a funnel-shaped net that trapped waterfowl after enticing them inside.[1]

"All writers agree—and there is no question—that the word *Decoy* came from Holland, where Decoys originated. It is an abbreviation of 'ende-kooy,' *i.e.*, in Dutch, the Duck Cage," wrote Ralph Payne-Gallway in his 1886 treatise *The Book of Duck Decoys*. "There is no doubt that at first the word 'kooy,' otherwise 'coy' (the cage), was used to represent the *cage of net* into which the fowl were driven in the early years of fowling, and in the later *enticed*."[2]

In its wider meaning, decoy has come to signify anything that lures or entices an entity away from its intended course, often into a trap. Biomedical scientists have employed the term to indicate various phenomena, such as a microbe whose presence interferes with the pathogenicity of another bug. However, biologists most commonly use the term decoy in reference to molecules that trap body messages, luring them away from the receptors that would deliver the information conveyed by those mediators to target cells. The first use of the word with this meaning was possibly in a 1962 article discussing mechanisms of protection against the influenza virus by way of a natural receptor decoy produced by fertilized chicken eggs. Most of the known molecular decoys are indeed soluble versions of the surface molecules that are necessary to transmit a mediator's message to the inside of a cell. It took three decades, though, for scientists to use consistently the term decoy in the realm of body message biology.[3]

In this chapter, we will review the process of discovery of molecular decoys that, by trapping cytokines, reduce the risk that powerful pro-inflammatory mediators might inflict harm to the body. We will see how researchers found cytokine decoys by deliberately searching for them, by looking for something similar, or by proceeding through a completely unrelated line of investigation. We will also follow one of these decoys as it developed into a popular drug that has changed the lives of millions of individuals, while making millions for the companies that sell it.

Luckily, tracing the history—and understanding the principles—of cytokine decoys is much easier that studying duck decoys, in large part because scientists are apparently more open than decoymen in sharing the secrets of their trade. "The history of Duck Decoys is surrounded by difficulties of no ordinary kind, for the art of constructing and working them was carefully concealed in former times," Payne-Gallway wrote. "A Decoyman lived a lonely, quiet life, seldom attending fairs or markets; he conversed in a low tone, and his appearance and manner was inoffensive and reserved: always as though going stealthily for fear of alarming the ducks."[4]

Writing in Color

Scientists can hunt for molecular decoys using many approaches, but one technique is queen among all: chromatography.

At the beginning of the twentieth century, Michail S. Tsvet, the botanist son of an Italian mother and a Russian father, came up with a new way of studying pigments, the colored substances that are present in plants. Because it separated green, orange, and yellow pigments—or perhaps thinking about the meaning of his own last name (*Tsvet* means "color" in Russian)—Tsvet called the new technique chromatography, that is, color writing.

Initially dismissed by Tsvet's contemporaries, chromatography has since become extremely popular among specialists of various disciplines, spawning several variants of the original version. Many of these variants have been decisive in identification of body messages, but it was a technique known as affinity chromatography that played the crucial role in the discovery of molecular decoys.[5]

The Art of Separation

Scientists use chromatography to separate individual molecules from mixtures made of disparate components, to get at a specific ingredient in a molecular soup. "Think of chromatography as a race," explains a Web site devoted to the topic. "Waiting on the starting line, you've got a mixture of chemicals in some unidentified liquid or gas, just like a load of runners all mixed up and bunched together. When a race starts, runners soon spread out because they have different abilities."[6]

For the kind of chromatography we are interested in here, the race starts at the top of a column, a long and thin cylinder made of glass or plastic and completely filled with tiny beads whose characteristics depend on the specific question being asked. The investigator applies a sample of the molecular soup to the top of the column. As the liquid falls down, pulled by gravity or by a pump, it passes by the beads, whose properties determine the speed at which the various ingredients in the soup will run. Those that run faster will get to the bottom quicker, the slow ones will take longer, while others may never make it down unless something forces them to leave the column.

Contingent on the investigator's specific question, one of many characteristics will determine the runners' speed down the beady column. If the goal it to separate molecules by their size, for instance, then size-exclusion chromatography is most helpful. This version of chromatography employs beads that are not solid but rather perforated by countless convoluted pores. Although small molecules can easily access the beads' pores, large ones cannot make it in there, they are excluded, hence the name of the technique. Therefore, small molecules have to travel a much longer route to get to the bottom of the column, diverted through all those labyrinthine pores, whereas large molecules go straight down in between the beads, avoiding the pores and thus quickly reaching the end of the column. Investigators select the size of the beads and the pores, as well as the length of the column, based on the level of size exclusion they need to achieve. By collecting the effluent from the bottom of the column in individual tubes—called fractions—the analyst can effectively separate the big from the small: large molecules arrive first, little ones later.

Testing those individual fractions in a bioassay will help determine the size of the molecule the investigator is hunting for. For example, when Dr. Dayer and his colleagues were attempting to isolate what would later be called IL-1Ra from the urine of febrile patients (as described in Chapter 3), they used size-exclusion chromatography, among other procedures, to determine the approximate size of the protein they were interested in. After applying urine samples to the top of their column, the researchers collected fractions of different molecular sizes, as determined by the speed by which they reached the bottom of the column, in separate tubes. They then mixed these fractions in various proportions with recombinant IL-1 and put the mix on top of cells that responded to IL-1. Only in the fractions where IL-1Ra, the IL-1 inhibitor, was present could the researchers observe suppression of IL-1 activity, thus allowing determination of the molecular size of their still-unknown mediator.[7]

Rather than selecting by size, affinity chromatography instead separates molecules based on their ability to attach strongly to other molecules, by their chemical affinity. Let's say you want to figure out if a sample contains a molecule that binds to a given cytokine. First, you need to have that cytokine in recombinant form and chemically attach it to your tiny beads, so that it covers the beads' surface like a molecular paint. Then, you fill your column with the cytokine-coated beads and apply the sample at the top. As the fluid percolates down the column, molecules that have an attraction for your cytokine—those that have high affinity—will attach to the cytokine-painted surface of the beads and get stuck there. After you've let your sample run through and rinsed the column to make sure nothing unwanted is left in there, you force your molecule to detach from the cytokine-coated beads by flooding the column with a (relatively) harsh chemical that overcomes the attraction of the two molecules for each other. While the cytokine remains stuck to the beads, the cytokine-binding molecule you are trying to isolate runs down to the bottom of the column, where you are ready to collect it in a tube and test for its presence with a bioassay.

While the principle of affinity chromatography is relatively easy to understand, its practice is rather arduous. To actually pull it off, a researcher needs to be aware of numerous subtle issues, optimize conditions for each specific study, and make sure everything works as it should, ready to deploy sophisticated tricks if and when things don't go as planned.[8]

Let's meet the scientist who mastered this technique, bending it to her will so it could be exploited for many a critical discovery.

The Artist

When it comes to cytokines and their receptors, the grand master of affinity chromatography is Dr. Daniela Novick, recently retired from the Department of Molecular Genetics at the Weizmann Institute of Science in Rehovot, Israel. During our interview Dr. Novick started out by sketching some of the major events in her life and that of her parents:

> I was born in Poland in 1948, a daughter to two Auschwitz survivors. In 1937, my father completed his medical studies in Paris, and then returned to Poland. During 1939–1945, he survived the Kielce ghetto, two and a half years in Auschwitz, a death march, and half a year in Mauthausen-Ebensee. His first wife and (apparently) two kids were murdered in Auschwitz. He was liberated in May 1945 by the United States Army. In Auschwitz, my father got to know the woman that became his second wife and my mother. She survived the Warsaw ghetto, one and a half years in Auschwitz, a death march, and finally was liberated by the Soviet Army in Ravensbrück.
>
> My parents were reunited immediately after the war, got married, and my sister and I were born. My father got a position as head of the Internal and Infectious Diseases Department in a hospital in Silesia (Poland). In 1957, the family immigrated to Israel and my parents started their life all over again.

I was almost ten years old. In Israel, I integrated in the fifth grade of elementary school and had to study two new languages, Hebrew and English. After high school, during the Six-Day War, I served in the Israeli Air Force.

My father had a wish that at least one of his daughters would study medicine. The compromise was that both my sister and I studied life sciences. Already in high school I liked biology, so the choice was only natural for me.

Immediately after the Six-Day War of 1967, I completed my Master in Microbiology at Tel-Aviv University, spent one year at the University of Cambridge, and in 1979 received my PhD from the Weizmann Institute of Science. Since then I have been a researcher at the Weizmann.

When deciding where to do my PhD, I realized that my tendency was to choose a mentor doing basic science but with a clear horizon in applied research, more specifically in the study of human diseases. Indeed, I chose a lab studying myasthenia gravis (an autoimmune disease) and was lucky to work out an experimental model in mice for this disease. After my PhD, I chose a lab—led by Prof. Michel Revel—studying interferon, a mediator that turned out to be a master cytokine in many diseases.

My father's wish that I become a medical doctor was not fulfilled in its conventional way. However, the dream of moving from bench to bedside became a reality when part of my work was translated into clinically useful therapies. One outcome of research to which I contributed is the drug Rebif™, used for treatment of multiple sclerosis, and the second is Enbrel™, used by hundreds of thousands of patients for treatment of rheumatoid arthritis. Looking back, I could not have asked for a better reward.

The repeated discoveries I made during my life would not be possible without a strong belief and conviction things would work, without daring, without an unlimited ambition to succeed where others had failed, without hard, sometimes dirty and sisyphic work, without curiosity and above all without passion.[9]

That passion and hard work allowed Dr. Novick to apply affinity chromatography to identification of several molecules, many of which act as cytokine decoys. The first such discovery, however, was not a decoy but the cell-surface receptor for the cytokine interferon-γ, which several investigators had been searching for a long time. In 1987, Drs. Daniela Novick, Menachem Rubinstein and their colleagues were one of the two teams that succeeded in identifying the receptor for that critical cytokine, and they did it via affinity chromatography. Since that first discovery, Dr. Novick's skillful handling and refined understanding of the color writing technique led to discovery of soluble receptors and binding proteins for an astonishing number of body messages, as she can claim rights to no less than ten such molecules, some of which we are going to explore in the following pages.[10]

Soluble TNF Receptors

Researchers who cloned the cytokine TNF in the mid-1980s wanted to identify the monocyte-derived mediator that killed cancer, as we saw in the previous chapter. Despite knowing that cytokines work by attaching to and activating specific receptors on the membrane of target cells, and despite early evidence for the existence of specific TNF binding sites on the surface of cells, for several years the suggestion kept floating around that TNF was endowed with intrinsic killing abilities that didn't require receptors. It took five years for scientists to identify the two receptors for TNF and demonstrate beyond doubt that this mediator acted just like any other cytokine, by binding to and activating receptors on target cells. Curiously, the discovery of molecules that inhibited TNF's activity, which turned out to be soluble versions of TNF receptors, preceded—and even enabled—identification of their cell-surface counterparts, while launching a new and highly consequential field of drug development.

The First Type

Three research teams went searching for TNF inhibitors at around the same time and all found what they were looking for in people's urine. Readers beware: despite having no choice but to present these findings in a sequence based on the date of first publication, the three teams acted simultaneously. "We knew about each other's work and we of course competed, although in a very friendly way," Dr. David Wallach, who directed one of the three teams, told me. Indeed, it was findings from his group, which comes last in the sequence I am going to present, that generated the patent for soluble TNF receptors. "In terms of patent we were first," Dr. Wallach explained, "but, scientifically speaking, we all did it at the same time."[11]

As previously reviewed, in the mid-1980s in Switzerland, Jean-Michel Dayer had demonstrated that urine collected from febrile patients contained IL-1Ra, an inhibitor of IL-1's activity. Given that TNF shared many pro-inflammatory properties with IL-1, Dr. Dayer reasoned an inhibitor of TNF might also exist and thus began searching for it. The investigators—Jean-Michel Dayer, Philippe Seckinger, and Sylvia Isaaz—started out with fifteen liters of urine pooled together from five patients with various inflammatory conditions. They concentrated their urine samples (by removing most of the water), applied size-exclusion chromatography (and additional methods) to separate the various molecular components, and collected fractions corresponding to proteins of various sizes. They then mixed each fraction with recombinant TNF and used a bioassay to test whether their samples might contain a molecule able to inhibit the activity of TNF, just as they had done a few years earlier when searching for the IL-1 inhibitor. In April 1988, the investigators published their first findings, demonstrating the presence in their patients' urine of a protein with the ability to inhibit the activity of TNF. The three authors concluded their article by correctly predicting that a whole family of molecules that interfere with the action of cytokines would likely exist.[12]

Meanwhile, size-exclusion chromatography and other versions of the color writing technique also allowed Dr. Inge Olsson's team at the University of Lund, in Sweden, to purify an inhibitor of TNF, this time from serum and urine of patients who were on dialysis due to chronic renal failure. These investigators took their findings one step forward, showing that the inhibitor physically interacted with TNF; based on this property they called the molecule TNF binding protein (TBP).

Proof of a direct interaction between TNF and TBP highly suggested that this new inhibitor would suppress the activity of TNF by a mechanism different from the one IL-1Ra used with IL-1. Whereas IL-1Ra blocked IL-1's biological function by attaching to the IL-1 receptor, thus preventing IL-1 itself from activating it, Dr. Olsson and his colleagues instead found that TBP bound to TNF itself, not to the TNF receptor.

Therefore, the TNF inhibitor did not act as a receptor antagonist that displaced the ligand from its receptor, but rather as a molecular decoy that diverted TNF away from the receptor present on the cell surface, trapping the cytokine into an unproductive interaction.[13]

In a follow-up article, the Olsson's group built an affinity chromatography column by coating beads with recombinant TNF and proceeded to purify and sequence TBP from two hundred liters of urine obtained from patients with kidney failure, basing the separation on TBP's ability to bind specifically to TNF (they exploited other techniques, too, and concentrated their two hundred liters of urine down to four).[14]

In Israel, Dr. Wallach, who—as we know—had been studying TNF for several years, also went looking for a TNF inhibitor. The researcher in charge of this project was PhD student Hartmut Engelmann, who combined various kinds of chromatography (but not affinity chromatography) into a complex four-step procedure that succeeded in isolating TBP. Here is how Dr. Wallach described the thought process behind those studies during our interview:

> I didn't start by looking for soluble receptors. In the initial stage of my studies, even before we isolated TNF, the discovery I take greatest pride in is the finding that if you take a preparation of what we called cytotoxin and put it on cells, you can see that it does not only kill cells but also induces resistance to killing. When we started studying the mechanism of how TNF induces resistance to its own function, we realized that there was more than one mechanism. The studies of Cerami highlighted that TNF could cause disease. Since in our studies of cell death we found that TNF induced mechanisms that antagonized its own action, I said—and here comes really a wishful thinking—I know several intracellular mechanisms by which TNF induces resistance to its own function. If there is an extracellular mechanism, if cells that are exposed to TNF secrete a TNF antagonist, then I will find it and it may be used as a drug against TNF. So, I thought, we will just take any liquid, any biological liquid of people that have TNF-related diseases. Initially, we tried to collect fluid from blisters, but that's too little material, so I thought maybe we could take serum, but there are so many proteins in serum: if there is anything it will be very difficult to isolate. Then, I thought, the kidney is doing the job, it cuts out albumin and various other proteins, so—it's all wishful thinking—if people who have TNF-related

diseases secrete in the blood something that would antagonize TNF, and if whatever is secreted is small enough to pass through the kidney, then we would find it in the urine and then it would be easy to isolate. And that's why we used urine.[15]

After using fluid from blisters and urine from sick patients, the Weizmann researchers eventually purified TNF binding proteins from urine collected at the men's lavatories in their department, and later switched to using samples of concentrated urinary proteins obtained from the pharmaceutical company Serono. Chromatographic approaches allowed them to isolate a protein that inhibited TNF's activity by binding directly to TNF, just like Dr. Olsson was finding at about the same time.[16]

Thus, in a very short period, all three research teams were able to confirm that the inhibitor they each had discovered—TBP—was the same protein and that it worked by binding directly to TNF. A year later, Dr. Wallach and his colleagues used TBP as a tool to clone one of the long-sought-after TNF receptors. Through these studies, they demonstrated that TBP was a soluble version, the extracellular portion, of one of the two cell surface TNF receptors.[17]

Evidence was emerging at the time from the laboratory of Dr. Carl Nathan at Weill Cornell Medical College in New York City, that activated neutrophils shed TNF receptors, turning the surface molecules into soluble versions that looked a lot like TBP. Partly based on these findings and on Dr. Nathan's own reasoning, Dr. Wallach and colleagues suggested that an enzyme might clip the part of the TNF receptor that dangles outside the cell and turn it into a soluble form, shedding it from the membrane. Generation of soluble receptors would therefore inhibit TNF's activity in two distinct ways: by shedding receptors from their surface, cells would reduce the intensity of TNF signaling, since the process would leave the cells with fewer binding sites available to be activated. At the same time, the clipping process would increase the number of soluble receptors acting as decoys, trapping TNF and preventing it from activating other cells. Indeed, subsequent studies by several investigators demonstrated the veracity of these insights. Scientists later came to recognize the wide applicability of the principle of receptor shedding, of enzymes that clip molecules away from the cell surface, turning them into floating decoys.[18]

The Second Type

If one urinary TNF inhibitor were not enough for the Weizmann researchers, another TNF binding protein was lurking, waiting to be discovered by the power of affinity chromatography, which a trio of scientists—Hartmut Engelmann, Daniela Novick, and David Wallach—now applied to the search. In a short memoir, Dr. Novick explained how, based on her experience in using affinity chromatography columns to identify the interferon-γ receptor in 1985–1986, she was confident the same technique would work with TNF. "Hartmut's skills, tremendous enthusiasm, and ambition, accompanied by the desire to design an alternative to the laborious purification procedure of TBP made him a perfect partner." They prepared a new affinity chromatography column that produced a big surprise, this time yielding not one, but two separate TNF binding proteins. While the first version was the same TBP previously isolated by Dayer, Olsson, and Wallach, the second one was brand new and rather unexpected.[19]

In January 1990, Drs. Engelmann, Novick, and Wallach reported isolation of the two types of TBP, suggesting these molecules might represent soluble versions of the two TNF surface receptors. A few months later, researchers at Synergen, the Colorado biotech that had just cloned IL-1Ra, also reported isolation of the second version of TBP, which they had isolated by affinity chromatography from supernatants of *in vitro* cultures of cell lines. Cloning of the protein confirmed identity of this second version of TBP with one of the two TNF receptors, the one called type 2.[20]

In just a few years, a modified version of this second TBP became the blockbuster drug Etanercept (Enbrel™), prescribed to millions of patients to treat rheumatoid arthritis, psoriasis, and other chronic inflammatory conditions.

One Cytokine, Two Drugs

In the 1970s and 1980s, research on rheumatoid arthritis was one of the three routes that had driven scientists to identification of IL-1, as we saw in Chapter 3. However, already in the mid-1980s and in parallel with their research on IL-1, Drs. Dayer, Beutler, and Cerami, and,

independently, Dr. Saklatvala demonstrated that the pro-inflammatory effects of TNF overlapped with those of IL-1 in the joints, suggesting that both cytokines might be important in the pathogenesis of
rheumatoid arthritis. Moreover, early studies by Dr. Dinarello and
others had shown that TNF induces IL-1, that is, that TNF sits
upstream of IL-1 in the inflammatory cascade, and that the two cytokines together are much more powerful than the sum of their activities, synergizing in provoking inflammation and tissue damage.[21]

By the late 1980s, evidence from various investigators—including
Drs. Marc Feldmann and Ravinder Maini in the United Kingdom—had
clearly indicated that the joints of rheumatoid arthritis patients also
contain elevated levels of TNF, not just IL-1. Importantly, Drs. Feldmann and Maini applied a subtraction approach to demonstrate that
interventions that blocked TNF's activity resulted in suppression of
IL-1 production, their data confirming that TNF was upstream of IL-1
in the cascade of mediators of the inflamed joint. Therefore, preventing
TNF from activating its receptor would kill the proverbial two inflammatory birds with a single molecular stone, directly neutralizing TNF's
activity and thus impeding TNF from inducing production of IL-1.[22]

In 1992, Drs. Feldmann and Maini launched the first trial of
anti-TNF therapy in patients with rheumatoid arthritis. To block TNF,
they used what would later become Infliximab, the anti-TNF antibody
developed in Dr. Jan Vilcek's laboratory that had been effective in
reducing intestinal inflammation in patients with IBD (see Chapter 4).
As explained in the description of the Albert Lasker Award for Clinical
Medical Research, bestowed upon Drs. Feldmann and Maini in 2003,
treated patients reported amazing relief only hours after infusion of
anti-TNF antibodies, feeling their joints had loosened up. In just a
few weeks, people who had been almost immobilized due to pain and
stiffness were able to play golf and climb stairs. In 1998, Infliximab—
originally approved for treatment of IBD—also became available to
rheumatoid arthritis patients and later to individuals with other
chronic inflammatory diseases such as psoriasis.[23]

In the meantime, the tools of molecular biology had allowed Bruce
Beutler, who had done pioneering studies on cachectin/TNF with Dr.
Cerami and had since moved to the University of Texas in Dallas, to
tweak the structure of TBP, generating a hybrid protein that was half

TBP and half antibody. Dr. Beutler had decided to generate this molecular chimera in a successful attempt to stabilize TBP, prolonging its persistence when administered *in vivo* and thus making it more suitable as a drug.[24] The rights to the technology for producing Beutler's hybrid molecules were licensed to the biotech Immunex (later acquired by Amgen), which employed this approach to generate Etanercept, a drug made by fusing part of an antibody to the second type of TBP, the soluble version of the TNF receptor Drs. Engelmann, Novick, and Wallach had discovered via affinity chromatography. Immunex soon brought Etanercept to the clinic for treatment of rheumatoid arthritis, the first molecule of such hybrid kind ever used in patients.[25]

Two different types of TNF-blocking medications therefore became available in the late 1990s: an antibody, Infliximab, and a chimeric soluble receptor, Etanercept. However, a quandary soon emerged. Whereas both Etanercept and Infliximab (and the other anti-TNF antibodies that were later developed) were effective in treatment of rheumatoid arthritis, only the antibody Infliximab, not the soluble receptor Etanercept, ameliorated disease in patients with gut inflammation caused by IBD. Yet both types of drugs attacked the same target, the cytokine TNF. What could possibly account for such a discrepancy?

A few years after approval of Infliximab for treatment of IBD, Dr. William Sandborn and his colleagues tested the effectiveness of Etanercept in IBD patients, probably fully expecting the new drug would work just like Infliximab. In a clinical trial published in 2001, they enrolled forty-three IBD patients, giving a placebo to half of them and treating the other half with Etanercept at the dose and schedule of administration that had already proved effective in rheumatoid arthritis. To everybody's surprise, the study came up completely negative, Etanercept having no effect whatsoever in IBD. The investigators concluded that either use of higher doses or more frequent dosing might be required to obtain a significant response in IBD patients.[26]

These additional studies, however, never took place. Evidence that Infliximab might do more than just block TNF's activity soon emerged, contributing to dampened enthusiasm for giving Etanercept another chance to work in IBD. In fact, studies demonstrated that Infliximab, but not Etanercept, promoted anti-inflammatory responses in the gut of IBD patients, from upregulation of T regulatory lymphocytes and fire extinguisher IL-10 to killing of dangerous immune cells. Later, investigators also demonstrated that enzymes present in the inflamed intestine of IBD patients inactivated Etanercept by cutting the hybrid molecule into its two pieces (the antibody part and the soluble receptor part), while Infliximab and the other anti-TNF antibodies remained active even after enzymes had cleaved them. Therefore, Etanercept is not used for treatment of IBD, even though clinical trials attempting to implement perhaps more appropriate doses—as suggested in 2001—never took place.[27]

However, lack of approval for IBD did not prevent Etanercept from joining Infliximab and another anti-TNF antibody, Adalimumab, at the very top of the list of top-selling drugs (as far as dollars go) for several years in a row. In 2014, Infliximab was fifth on the list of income-generating drugs, grossing $8.5 billion, Etanercept was third at $9.2 billion, while Adalimumab won the race at $12.5 billion. To understand these bewildering figures we should add up three factors. The first is the ever-increasing number of individuals affected by the chronic inflammatory disorders that can be alleviated by anti-TNF medications. The second is the chronic nature of such disorders, which can be treated, but not cured, by blocking TNF. Once initiated, individuals generally use anti-TNF compounds for several years, even decades, thus piling up prescription after prescription. Lastly, the drugs come with a hefty price tag—in the tens of thousands of dollars per person per year—a price that is only partly justified by elevated production costs. While contributing to income for the companies that make the drugs, these prices obviously create insurmountable barriers to access for the many individuals and health care systems that cannot even begin to afford such expensive treatments.[28]

When considering patterns of utilization and prices of a medication, it is always interesting to look at the process that led to the drug's discovery and development, as we just did for Etanercept. It is also informative to identify the original source of the biological material that made that discovery possible.

Drugs from Nuns

A few pages back, you might have noticed the name of a pharmaceutical company, Serono, in the account of Dr. Wallach's isolation of

TBP. While Jean-Michel Dayer and Inge Olsson had found TBP in urine of patients affected by various diseases, Dr. Wallach first collected samples in his department's men's lavatory; realizing larger quantities were necessary, he later turned to an external source, the pharmaceutical company Serono (currently part of Merck). Drs. Engelmann, Novick, and Wallach then utilized the same source of urine for purification of the second version of TBP, the one that was turned into Etanercept.

Cesare Serono founded his pharmaceutical company in Rome, Italy, in 1906. For many years, the company mostly produced compounds isolated from chicken eggs. In 1949, the chemist Piero Donini developed an effective technique for extraction of gonadotropins, hormones that promote egg and sperm production, from urine. This technique allowed Serono to formulate Pergonal, an injectable drug made of gonadotropins isolated from urine, which turned the company into the worldwide leader for treatment of infertility. Synthesized and released by the pituitary gland (and the placenta during pregnancy), gonadotropins can do several things, including stimulating egg maturation in women's ovaries. Drugs containing gonadotropins, like Pergonal, are widely used to treat infertility and to induce ovulation prior to egg collection for *in vitro* fertilization procedures. Now largely replaced by recombinant proteins, for decades Serono made these drugs by extracting hormones from urine using Donini's technique. The urine of postmenopausal women is particularly rich in gonadotropins, whose production highly increases after the ovaries cease to function.[29]

In addition to Donini, another scientist was instrumental in forging Serono's fortunes in the treatment of infertility, the Austrian-Israeli reproductive endocrinologist Bruno Lunenfeld. In the late 1950s, after having obtained an MD and studied fertility hormones during his PhD studies, Dr. Lunenfeld wanted to take gonadotropins to the clinic. In 1957, Lunenfeld contacted Piero Donini at Serono to discuss his plans. However, Lunenfeld and Donini initially had trouble convincing Serono's board of directors of the project's feasibility, since collection of vast amounts of urine from postmenopausal women was required. At the time, the Vatican controlled a major share of Serono. Representing the Vatican on the company's board of directors was Don Giulio Pacelli, a nephew of the pope. After many discussions, Lunenfeld and

Donini managed to persuade Don Pacelli of the importance of gonadotropins for treatment of infertility, an issue the Catholic Church was
definitely concerned about. Eventually, Don Pacelli convinced Serono's
board to go ahead with Lunenfeld plans by declaring that nuns housed
in old-age homes would provide the required urine.[30]

Because it takes about a day's output of urine from ten women to
produce a single dose of Pergonal, postmenopausal nuns—initially at
convents in Italy and later also in other countries—for many years collected a mind-boggling amount of urine that went into production of
the fertility drug. After isolation of gonadotropins, though, Serono had
no use for the leftovers of that process, which for decades went to the
dumpster.

Working in Israel, Dr. Lunenfeld, and thus the company Serono,
was closely associated with the Weizmann Institute. The company
was involved in cytokine research at the Weizmann, particularly with
research by Dr. Michel Revel (Daniela Novick's postdoc advisor) on
the cytokine that later became the drug Rebif™, developed and commercialized by Serono. Moreover, in the 1980s Serono supported
some of David Wallach's research on TNF. "Initially we took clinical samples of people with diseases, but then we found TBP also
in normal urine," Dr. Wallach recalled during our interview, referring to the initial collections in the men's lavatories. "We actually
didn't realize it initially, but the company that happened to support us
at the time, Serono, made its fortune because of work in urine. We
thought—wow—maybe we can convince them to send us urine. That
was a technical breakthrough, because they sent us concentrates of
urine from nuns."[31]

And so it was that Catholic nuns were the source of two medications used by millions of people worldwide, Pergonal for infertility,
directly extracted from the nuns' urine, and Etanercept, based on the
TNF decoy Drs. Engelmann, Novick, and Wallach discovered by
pouring the leftovers of Pergonal's production on top of their affinity
chromatography column. Whether the nuns ever consented to be part
of this process, I could not ascertain.[32]

As productive as it has been, the direct route of affinity chromatography analysis of urine has not been the sole conduit to cytokine
decoys, as shown by the next molecule, the decoy of pro-inflammatory

IL-1, discovered through a completely different process and following a much more circuitous path.

The IL-1 Decoy

Either by tinkering—as François Jacob would say—or through some sort of logic, evolution is definitely creative in its ability to come up with ever-novel solutions to the very same problems. An example of such evolutionary creativity is the decoy of the cytokine IL-1, similar and yet quite different from those of its sister pro-inflammatory cytokine TNF.

We already discussed in some detail the discovery and biological activities of IL-1Ra, the antagonist of IL-1 that has no counterpart that we know of in the TNF system. Besides IL-1Ra, IL-1 also has a receptor-like decoy that, like those of TNF, traps the cytokine and diverts it from activating target cells. However, whereas TNF's receptors act as decoys only in their soluble form, both the membrane-bound and the soluble version of one of the two IL-1 receptors, the one known as type 2, can intercept IL-1's message.

It is in thinking about this molecule's function that Dr. Alberto Mantovani, leader of the team that discovered the IL-1-inhibiting properties of membrane-bound IL-1 receptor type 2, came up with the designation of decoy. "I remember perfectly well how I thought about the word *decoy* one night," Dr. Mantovani told me. "The next morning I went to the lab and everybody liked the name. The term then came to be used to define a general strategy for regulation of cytokines and other mediators." During our interview, Dr. Mantovani also explained how he ended up working on IL-1 by deciding not to be a physicist and by not speaking German.[33]

How to Become an Immunologist

"In high school, I was very good in mathematics and physics," Dr. Mantovani began. "My professor at Liceo Manzoni, in Milano, [Italy] was Umberto Forti, a physicist who had been part of Enrico Fermi's circle in Rome. Professor Forti was very happy when I told him of my decision to become a physicist." It was the spring of 1967 and young

Alberto enrolled at the faculty of physics at the University of Pavia, near Milano. That same summer, however, he took off for the United Kingdom as part of a volunteer camp organized by a Quaker pacifist organization, working in a psychiatric hospital in Oxford. "We took psychiatric patients on vacation," Mantovani explained. That experience had him backpedal on his career goals: he would become a physician, not a physicist. Indeed, upon returning to Italy in the fall, he enrolled in medical school at the University of Milano.[34]

During his second year as a medical student, Mantovani decided to try out research, applying to the equivalent of an MD/PhD program. "I applied to the Institute of General Pathology and my great fortune was to not speak German," Dr. Mantovani elaborated during the interview. "The director of the institute was an old pathologist and at that time German was still the language of choice for old pathologists. The only student who spoke German was considered the best one and assigned to the old pathologist. But that was not a good lab. With my knowledge of English and French, I arrived second, happening upon a great research group directed by Dr. [Guido] Guidotti, who had one of the very rare NIH grants assigned to investigators outside the United States and was a pioneer in the application of molecular biology techniques to study protein metabolism. I was lucky enough to be in a brood of people at different stages of training, and we all ended up doing very well." During medical school, Mantovani also attended an immunology course taught by an excellent professor, a class where students were exposed to both hands-on experiences and seminars by international speakers.[35]

During his fifth year in medical school, needing to find a more consistent source of income than the odd jobs he had taken until then, Mantovani asked his mentor about fellowship opportunities. Since none were available at the university, a meeting with Drs. Silvio Garattini and Emilio Mussini, two of the founders of the Mario Negri Institute for Pharmacological Research, was arranged. Established as the legacy of jeweler and philanthropist Mario Negri, the institute had opened its doors in 1963 in the periphery of Milano, the first private Italian foundation dedicated entirely to biomedical research; it soon grew to be one of Italy's premier research centers. After a challenging interview with the two founders, Mantovani was granted a fellowship at the

Mario Negri Institute and began to work under the supervision of Dr. Federico Spreafico, a tumor immunologist who at the time directed the immunology section.[36]

After graduating from medical school, Dr. Mantovani decided to specialize in oncology and started training in the pediatric ward of the Cancer Institute in Milano while continuing his experimental studies at the Mario Negri. "That's when I decided I wanted to do research, not clinical work," he clarified. To acquire international exposure, Mantovani chose the laboratory of Dr. Peter Alexander, a tumor immunologist specializing in macrophage biology at the Chester Beatty Research Institute in England. "The project I was assigned did not work," Dr. Mantovani explained. "So I came up with my own experimental questions. Recognizing the project was solely the fruit of my ideas and practical skills, my mentor let me publish the results on my own, resulting in three solo publications I am still very fond of."

At the end of that experience in the United Kingdom, in the late 1970s Mantovani returned to the Mario Negri Institute, where he developed his own line of research by blending tumor immunology with macrophage biology and the nascent field of inflammatory cytokines. He remained at the Mario Negri for about three decades, before moving to his current position as research director of the Humanitas Clinical and Research Center, near Milano.

Converging Parallels

Two independent areas of research in Dr. Mantovani's laboratory at the Mario Negri permitted identification of the decoy properties of IL-1 receptor type 2, albeit neither started out by asking anything close to the final discovery.

The first area of investigation emerged from a collaboration with Dr. Elisabetta Dejana, another researcher at the Mario Negri Institute. The inner lining of blood vessels, the endothelium, is Dr. Dejana's major interest. The endothelium is made of long and thin cells, assembled in a single layer that completely covers the inside of arteries, capillaries, and veins. These cells are not just a passive covering that separates blood from tissues but instead actively participate in coordinating cardiovascular function, blood clotting, and inflammatory responses.

The endothelium controls vessels' contraction and relaxation, thus playing an essential role in the regulation of blood pressure. Alterations in endothelial cells are also among the very first events in formation of atherosclerotic plaques in the arterial wall and in stimulation of blood clotting when vessels are damaged. Moreover, it is the endothelium that exposes on its surface the chemical Velcro that allows leukocytes to stop and move from blood to tissues in inflamed areas.

Part of classroom teaching for what seems like a very long time, the role of inflammation in regulating the function of endothelial cells had received only limited attention until the mid-1980s. In 1984—the year IL-1 was cloned—Drs. Dejana and Mantovani began their collaboration with a seminal article demonstrating that inflammatory mediators produced by activated leukocytes impel the endothelium to generate high levels of prostacyclin, a molecule that is essential in controlling cardiovascular function and that looks similar to the prostaglandins that cause fever. The investigators first suggested, and later confirmed, that pro-inflammatory IL1 was possibly the most potent cytokine in upregulating prostacyclin production by the endothelium.[37]

While the collaboration on IL-1 and endothelial cells proceeded, with numerous additional important findings, Dr. Mantovani also pursued projects related to his other major interest, tumor immunology.

There are two pieces to this other area of investigation. The first piece is a family of genes called proto-oncogenes, which regulate cell growth and differentiation. When proto-oncogenes mutate, which can happen for many reasons, they lose the proto part of their name and become oncogenes, an important cause of cancer. Recognition that tumorigenic oncogenes are nothing but mutated versions of normal genes came in the mid-1970s and was at the origin of the 1989 Nobel Prize in Physiology or Medicine to Drs. J. Michael Bishop and Harold Varmus.

Neutrophils, the leukocytes that arrive first at the site of inflammation, are the second piece of the investigation. Neutrophils live a very short life, a few days at most. The general assumption in the 1980s was that neutrophils' life span was genetically predetermined, the environment playing no significant role in how long these cells lived.

Combining the proto-oncogene and the neutrophil pieces, Mantovani's team demonstrated that neutrophils contained high levels of a proto-oncogene that is important in cell survival, and that exposure to inflammatory stimuli packed these leukocytes with even more proto-oncogene. These results highly suggested that life span in neutrophils was not necessarily genetically predetermined, that their longevity could be altered by environmental factors through modulation of proto-oncogene levels.[38]

A few years later, in 1992, Dr. Mantovani's group returned to this concept, bringing together research that had started as a study of IL-1's effects on the endothelium with investigations on proto-oncogenes and neutrophils' survival. In joining the two areas, the scientists demonstrated that, by interfering with the genetic program of cell death, exposure to IL-1 dramatically prolonged neutrophils' life span in the test tube, from an average of 35 hours to a whopping 115.[39] These were critical findings, showing that inflammation could make neutrophils live a much longer life than expected, that the microenvironment these leukocytes inhabited contributed to determining how long these cells stuck around. The data had important implications for a better understanding not only of inflammation but also of the respective roles of genetics and the environment in determining a cell's life span. However, nothing resembling research on a molecular decoy was even remotely in sight.

The Receptors for IL-1

In the early-1980s, even before cloning of IL-1, investigators had already started to search for this cytokine's receptors. Many scientists participated in the quest, but those at the biotech Immunex were the first to succeed in cloning the IL-1 receptor genes.

In 1988, Dr. John Sims and his colleagues at Immunex reported the genetic sequence of a receptor that bound both IL-1α and IL-1β. In their *Science* article, the investigators mentioned that an additional version of the IL-1 receptor might exist, as other researchers had noted. In 1991, Immunex scientists found that second receptor, too. This one had a peculiarity, though: its intracellular tail was very short, barely reaching inside the cytoplasm.

The function of this second IL-1 receptor remained unclear, the article's authors being unsure as to whether the signaling pathways used by the two IL-1 receptors partly overlapped or were completely independent. In about a year, in 1992, understanding of intracellular pathways activated by IL-1 receptors made a giant leap with demonstration by Luke O'Neill, Jeremy Saklatvala and colleagues that IL-1 receptors type 1 stimulated the NF-κB signaling pathway—the same pathway activated by TNF—while type 2 receptors did not.[40]

Meanwhile, on the wave of emerging information on soluble cytokine receptors and antagonists, in 1990, Dr. Gordon Duff and his colleagues at the University of Edinburgh, in the United Kingdom, and Drs. Judith Giri and Richard Horuk, at DuPont de Nemours, in Pennsylvania, published a series of reports demonstrating that biological fluids and cell culture supernatants contained an IL-1β binding protein that looked similar to the extracellular portion of an IL-1 receptor, probably type 2.[41]

To summarize, by the end of 1992 scientists had identified both of the cell-surface receptors for IL-1, simply called type 1 and type 2. Researchers knew that binding of IL-1α and IL-1β to type 1 receptor activated the NF-κB signaling pathway in the target cell, whereas binding of IL1Ra did not send an activating signal at all, actually preventing stimulation by both IL-1α and IL-1β. Knowledge about the type 2 receptor was much sketchier: all that investigators knew in 1992 was that type 2 receptor had a short intracellular tail and did not activate NF-κB. Also, there were soluble IL-1 binding proteins that could inhibit IL-1β's activity and that looked like soluble type 2 receptors.

The IL-1 Decoy Is Born

At this point—it's 1993—evidence from Dr. Mantovani's research team, working in collaboration with John Sims and colleagues at Immunex, rapidly unraveled the knots. First came demonstration that type 1 receptors mediated all of the cell-activating properties of IL1. Indeed, blocking type 1 receptors with specific antibodies made cells completely deaf to IL-1. It didn't matter whether type 2 receptors were present or not, when scientists disabled type 1 receptors IL-1 simply could no longer deliver its message.

Equally important was the observation that cells could still clearly hear IL-1's message when type 2 receptors were incapacitated. In truth, that message became even louder, as if type 2 receptors were somehow reducing its potency. In fact, the investigators found that, rather than blocking responses to IL-1, anti-type 2 receptor antibodies enhanced them. Based on observations by Drs. Duff and Giri and their colleagues that soluble IL-1 binding proteins likely were fragments of the type 2 IL-1 receptor, the Mantovani team speculated that the membrane form of this receptor served as the precursor for the soluble receptor, which would then act as an antagonist of IL-1. The researchers were getting close, but still suggested that type 2 receptors needed to be shed from the membrane to inhibit IL-1. A sensible proposal given the state of the knowledge at the time, with each established cytokine inhibitor being a soluble molecule.[42]

Finally, it was research that had started with a completely different goal—understanding the mechanisms that regulate neutrophils' life expectancy—that unveiled the true nature of IL-1 receptor type 2. Following up on their prior observation that IL-1 markedly prolonged the life span of neutrophils, Drs. Francesco Colotta, Fabio Re, and colleagues in Mantovani's group started out by showing that two separate anti-inflammatory mediators, the cytokine IL-4 and glucocorticoids, abolished this response. In fact, when the researchers exposed neutrophils to an anti-inflammatory environment and then added IL-1, the cells died just as quickly as they would have had they never been in contact with IL-1. This meant that the presence of anti-inflammatory mediators canceled out the ability of IL-1 to deliver its lifesaving message to neutrophils.

How did this happen? The investigators found that IL-4 and glucocorticoids both sharply increased the number of IL-1 receptor type 2 on neutrophils' surface. It was this increased density of type 2 receptors that drowned out IL-1's message. The anti-inflammatory environment could also induce shedding of type 2 receptors from neutrophils' surface, turning them into soluble molecules, but this was not strictly necessary for the IL-1 inhibitory effect to take place. Thus, both the membrane form and the soluble version of IL-1 receptor type 2 behaved as inhibitors of IL-1's activity, a whole new concept in the biology of body messages.

Additional studies in other cell types completed the report, which concluded by suggesting that the term receptor might be inappropriate for IL-1 receptor type 2, since this molecule did not transmit any signal inside the cell even when the cytokine was attached to it. The group proposed instead to refer to IL-1 receptor type 2 as a decoy, thus officially introducing this terminology into the field of cytokine biology. Subsequent studies by Gordon Duff and others worked out the details of binding of each ligand—IL-1α, IL-1β, and IL-1Ra—to membrane-bound and soluble forms of the type 2 receptor, while plenty of investigators picked up the topic and contributed to demonstrate the importance of this mechanism in controlling IL-1's inflammatory effects both *in vitro* and *in vivo*.[43]

To recap, TNF receptors convey that cytokine's message to the receiver cell while present on the cell membrane, becoming inhibitors only after enzymes clip them away, shedding the receptors in soluble forms into the surrounding fluids. On the other hand, IL-1 receptor type 2 does not signal even when attached to the cell, always acting as a decoy irrespective of its location. As uncommon as it might be, this behavior is not unique.

In the early 1990s, in fact, again using human urine together with refined versions of affinity chromatography, Daniela Novick and her colleagues identified both the soluble form and the cell-surface counterpart of the receptor for type I interferon (not to be confused with interferon-γ, also known as type II interferon), opening the way to understanding of this receptor's distinct signaling properties. Remarkably, type I interferons, cytokines with potent antiviral activities, were discovered in the late 1950s—the very first cytokines out there—but identification of their receptor, both soluble and membrane-bound, took almost forty years, until color writing came to the rescue. "There are very few eureka-like moments in one's career. This was one of them," Daniela Novick wrote commenting on her discovery.[44]

One version of the surface receptor for type I interferon has a short intracellular tail, just like IL-1 receptor type 2. In 1997, the team of Dr. Ed Croze at the University of Tennessee demonstrated the decoy properties of this short type I interferon receptor, thus providing a companion to IL-1 receptor type 2.[45]

Since the 1993 article reporting on the IL-1-inhibiting properties of IL-1 receptor type 2, the term decoy has come to designate a molecule that traps a ligand, diverting it from activating a cell. Even though the vast majority of known molecular decoys are soluble receptors, some— like IL-1 receptor type 2 and the short version of the receptor for type I interferons—exert their diverting function even when present on the cell surface. Because one of the postulates of receptor theory, as formulated at the beginning of the 1900s, states that some early chemical event must occur once a ligand binds to a receptor, whether the term receptor is even appropriate for molecules like IL-1 receptor type 2 that never even get the opportunity to signal remains questionable. Luckily, uncertain terminology is not an issue for our next decoy, IL-18 binding protein, a molecule that nobody ever mistook for a cytokine receptor.

IL-18 Binding Protein

Discovered in the late 1980s for its ability to increase production of interferon-γ, cloned in 1995, IL-18 is one of the members of the IL-1 family we met in Chapter 3.

As soon as information about the newly cloned cytokine was out, scientists went combing for its receptor. Daniela Novick was among them. Whereas other researchers resorted to traditional techniques involving binding of IL-18 to the membrane of cells cultured *in vitro*, Dr. Novick ran for her nuns' urine concentrates. After all, by that time she and her colleagues had already successfully isolated several soluble cytokine receptors from urine, and used some of them to identify surface receptors. Introduced in the late 1980s, the concept that most cytokine receptors exist in both a membrane-bound and a soluble form had been thoroughly embraced by the mid-1990s. It thus seemed altogether reasonable to search for the IL-18 receptor in urine, and Dr. Novick—as the grand master of affinity chromatography—was perfectly positioned to perform such studies.

Meanwhile, at the University of Colorado in Denver, Charles Dinarello had taken a keen interest in IL-18 due to this cytokine's similarity with IL-1β, which he had cloned a decade earlier (Fernando Bazan had proposed the name IL-1γ for IL-18, if you recall). Being close friends,

Charles and Daniela began to collaborate in the search for the IL-18 receptor, eventually involving other laboratory members, including myself. First at the Weizmann Institute with Dr. Novick and then joining our group in Denver, Dr. Soo-Hyun Kim was an essential contributor to this work, as was Dr. Menachem Rubinstein, senior author on most of Dr. Novick's reports.[46]

"In an effort to isolate a soluble IL-18 receptor, we passed a preparation of proteins concentrated from five hundred liters of normal human urine on an IL-18-agarose column," begins the results section of the article describing the use of affinity chromatography in the isolation of IL-18 binding protein (IL-18BP) from nuns' urine.[47]

That protein, however, wasn't even close to resembling an IL-18 receptor. In recalling the events leading to identification of IL-18BP, Daniela Novick described how on Yom Kippur (the Jewish Day of Atonement) in 1997 she "sinned" by opening her computer, only to learn that another group had succeeded in identifying the molecule she and Charles Dinarello were looking for, the receptor for IL-18. "I read the abstract of that paper and for the first time in my career I understood what it means to be scooped. It was of course a sleepless night until we compared the sequence of our protein to the one published. Luckily, our protein turned out to be different," she wrote in a short memoir.[48]

While Drs. Novick and Kim were attempting to isolate the IL-18 receptor in soluble form via affinity chromatography, other researchers had found that receptor on the surface of cells, publishing their results in 1997 and 1998. The IL-18 receptor belonged to the IL-1 receptor family, not a huge surprise given the similarities between IL-18 and IL-1, but still a fundamental addition to a more complete understanding of IL-18's biology.[49]

Something else was very new, however, because the molecule Daniela Novick and Soo-Hyun Kim had fished out—IL-18BP—was not a soluble version of the IL-18 receptor but a molecule in its own league. Its closest human relative was the IL-1 receptor type 2, but even that molecule wasn't a very good match. Crucially, IL-18BP acted as an IL-18 decoy both *in vitro* and *in vivo*, as demonstrated in our 1999 article published in the journal *Immunity*.[50]

Since then, work by many investigators helped define the biochemical properties of ILI8BP's multiple versions and evaluate the molecule's biology in countless manners. A pure decoy that exists only in soluble form, with a stupendous power to prevent IL-18 from delivering its message, IL-18BP is now considered as a potential therapeutic tool for several diseases, both common and rare. Indeed, in 2015 administration of IL-18BP saved the life of a child affected by an extremely rare type of autoinflammatory syndrome characterized by the presence of high levels of IL-18 in the bloodstream.[51]

Although IL-18BP was not what Drs. Novick and Dinarello had originally planned to isolate—they had started out looking for the IL-18 receptor, after all—this molecule nevertheless fit into the by-then widely accepted framework of cytokine decoys that Dr. Novick herself had contributed to generate. The discovery of IL-18BP, therefore, did not break any dogma, but rather enriched the established paradigm with a new, incredibly interesting, twist: a soluble decoy with no resemblance to a cell surface receptor. Yet, after describing how long and how hard Soo-Hyun Kim had searched for IL-18BP's relatives, in her too-short memoir Dr. Novick recounted her reaction to the new decoy, giving us a sense of a researcher's feelings when encountering something that doesn't fit into neat categories, "We hesitated at that time to state with confidence that our protein is an IL-18 binding protein rather than a classical receptor and therefore made no effort to publish it as such. We probably feared then to be included in the wrong category of scientists since 'If you have the same ideas as everybody else but have them one week earlier than everyone else, then you will be hailed as a visionary. But if you have them a few years earlier you will be named a lunatic.'"[52]

However, albeit in a league of its own, IL-18BP did actually resemble something, as did ILI receptor type 2, interferon receptors and TNF binding proteins. They all looked like molecules made by viruses.

Smart Viruses

Neither dead nor alive—or perhaps both dead and alive—viruses are sort of zombies that depend on live cells for reproduction. Viruses can be defined as noncellular entities consisting of genetic information, which can take the form of either DNA or RNA, that travels from one cell to another. A protein coat envelops that genetic information, allowing viral particles to gain access to a cell's inside. Once the viral particle is inside the cell, the cellular reproductive machinery makes copies of viral genes and churns out viral proteins by dipping into the cell's own store of building blocks.

An endless source of wonder, speculation, and investigation for scientists and philosophers alike, viruses are everywhere: in our cells, food, and waste; in bacteria, animals, and plants; buried deep inside glaciers, and filling the oceans. Scholars continue to debate the evolutionary origins of viruses, but there is no doubt these entities play an essential role in life as we know it.[53]

Like most bacteria, the vast majority of viruses are not pathogenic. Yet, we tend to associate viruses with infectious diseases: the flu and the cold, chickenpox, measles and mumps, hepatitis, AIDS. When pathogenic viruses enter the body, they are perceived as danger by the immune system through activation of pattern recognition receptors (remember the PRR of Chapter 1?). As soon as PRR on sentinel cells sense danger, they induce those sentinels to start producing body messages, including cytokines and chemokines. This causes inflammation, which eventually activates specific immune responses that involve participation of lymphocytes. If everything goes well, the infection is cleared. Sometimes, however, the immune system enters into a standstill with the virus, resulting in chronic infections that may go on for a lifetime or end with either of the two entities—the body or the virus—prevailing over the other.

Viruses are burglars. As they replicate inside the cells they have infected, they steal pieces of genetic material and make them their own (some scientists think this is how viruses came to exist, others disagree). When Dr. Soo-Hyun Kim was frantically searching databases for similarities between IL-18BP and known proteins, he did not find any good match in humans or other animals, but got much luckier when searching viral genomes. Indeed, the closest relatives to human IL-18BP were proteins encoded by viral genes. Several kinds of poxviruses (the most famous member of which is the virus that caused smallpox; yes, we can use the past tense here!) contain genes, in all likelihood stolen from cells, that code for proteins that look very similar to IL-18BP. These viral decoys are extremely effective at silencing IL-18 produced by the body's own cells, thus blocking this cytokine's ability to induce production of interferon-γ. Because interferon-γ is critical in setting in motion mechanisms that eventually get rid of the viral infection, by carrying a gene for its own IL-18BP the virus takes control of the immune response, swaying it to its own advantage.

This clever trick is in no way limited to IL-18. Viruses have stolen the genes for IL-1 receptor type 2 as well as the receptors for TNF, type I and type II interferons, and various chemokines, even coming into possession of their own version of anti-inflammatory IL-10 and other cytokines. The viral variants of these proteins mold the immune response to the viruses' advantage. Ironically, in doing so viruses force our cells to use their own resources for production of mediators derived from those long-ago hijacked genes, mediators that now work for the abductors. Ergo, it is a dynamic exchange of information that determines the outcome of a viral infection, a conversation that combines messages coming from the body with those encoded by viral genes (all produced by the body's cells using the cells' own building blocks and energy). These body and viral messages physically interact with each other, creating a whole new emerging meaning.[54]

It has been said that viral genomes are among the best immunology textbooks ever written. In writing their own immunology manuals, though, viruses not only heavily plagiarize from available content, they also change the spirit of the text, from an essay about a war against a foreign enemy to a treatise on diplomatic negotiations in which all parties contribute their viewpoints. But biology, like diplomacy, is full of nuances. Where deception and diversion will serve one party, encouragement or imposition may instead be the best tools under different circumstances, as is the case for the protagonists of the next chapter, interleukin-6 and its soluble receptor, which looks like a decoy, but a decoy it ain't.

6

Between Inflammation and Metabolism:
The Acute-Phase Response

OH, THE LIVER! It works like a horse and is stronger than a mule, yet
is better than a chameleon in adapting to changing environments. Like
a lizard's tail it even grows back when cut into pieces. When it fails,
life is gone.

Defined by most dictionaries as the biggest gland of the body, the
liver is an organ that defies easy characterization. For a long time the
liver puzzled physicians and scientists: Is it there to warm the stomach
or to generate blood? Is it the convergence of all veins? The liver is the
largest organ in the body, but what does it do?

When I teach students about the liver, I always mention the
often-repeated statement that this organ performs more than five hundred different functions. Now, five hundred is a big number and it is
meant to be just that, an indication that the liver is responsible for so
many things: from storing and activating vitamins to controlling iron
and sugar concentrations, from producing clotting factors to metabo-
lizing drugs and alcohol. The liver regulates blood pressure and
immune responses, makes and breaks proteins and fat, prevents skin
and eyes from turning yellow (by excreting bilirubin derived from old
red blood cells) and the brain from going into a coma (by removing

toxins). And much, much more. So much more that we cannot live without a liver. How can we possibly have a single definition for such an evolutionary marvel?

For centuries, Western medical tradition considered the liver (and its associated gallbladder) as one of the main sites of origin of the body's humors, the four substances whose balance was thought to be essential for maintaining not only physical but also mental health. Even though humoralism is no longer an accepted theory for the body's workings, the liver still holds a central place in our understanding of health and disease, in large part as responder and producer of body messages. In fact, a myriad of messages converge on the liver, while the liver coordinates an outgoing response that is itself made of countless mediators.

Widely considered the body's metabolic hub, scientists have known for decades that the liver is also a crucial participant in the response to infection and inflammation. Examining some aspects of this history will allow us to appreciate the manifold role of the liver and of its messages, as we slowly drift from inflammation to metabolism.

The Acute-Phase Response: Take One

Evidence from the early 1930s indicated the presence of a somewhat mysterious substance in the blood of patients acutely ill with pneumonia caused by a type of bacteria known as pneumococci. This substance suddenly appeared and reached maximal levels in the bloodstream during what was, and still is, called the acute phase of the disease, when the patient is usually feverish; it then disappeared almost as rapidly as it had shown up once the recovery phase began. In the 1940s, scientists found out that the mystery compound was a protein made by the body's own cells, a protein that received the unimaginative name of C-reactive protein, CRP in short, because it reacted with a sugar called C polysaccharide produced by the pneumococci. Quite surprisingly, scientists also noted a steep elevation in CRP levels in the blood of patients and experimental animals that were not infected with pneumococci but that were instead sick due to infections caused by completely different microbes, microbes that did not produce the sugar C polysaccharide to which CRP attached to. Even inflammation of noninfectious origin could raise CRP to sky-high levels. What could this mean, and where was this protein coming from? It took until the mid-1960s to point to the liver as the organ that produces and pumps out CRP in response to infection and inflammation.[1]

Meanwhile, data showing that CRP's behavior was not so unique began accumulating. Indeed, scientists demonstrated that a whole group of proteins produced by the liver—collectively called acute-phase proteins—dramatically increased during infection and inflammation. In fact, the blood of acutely ill patients contained greater than thousandfold higher levels of acute-phase proteins compared to blood of healthy individuals, implying that the liver worked like crazy to suddenly start producing enormous amounts of proteins, and then rapidly stopped as soon as the acute phase was over.[2] How did this happen? And why? The first question propelled researchers to the discovery of interleukin-6, a versatile body message that regulates both inflammation and metabolism. Getting an answer to the second question turned out to be much trickier.

Interleukin-6

Trying to understand the mechanisms by which infection and inflammation generated such a surge in acute-phase proteins, in 1945 Freddy Homburger at Yale University showed that injection into dogs of a protein-like substance isolated from abscesses (which are mostly made of leukocytes and dead bacteria) induced synthesis of an acute-phase protein known to be produced by the liver. This was possibly the first demonstration that a body message present at a site of inflammation directly affected protein production by cells of the liver, the hepatocytes. This theme was revived and refined in the 60s, 70s, and 80s, decades that, as we have seen in previous chapters, were buzzing with activity directed at identifying mediators of fever and inflammation. Researchers could now demonstrate that molecules produced by activated macrophages and other leukocytes increased output of acute-phase proteins: IL-1 and TNF were connected to the hepatic response almost as soon as these cytokines were discovered, but data indicated there must be more to the story. Using an *in vitro* assay they had developed a couple of years earlier, in 1983 David Ritchie and Gerald Fuller at the University of Texas partially characterized what they named hepatocyte-stimulating factor: a monocyte-derived protein that directly induced production of acute-phase proteins in hepatocytes but that, given its size and biochemical properties, was unquestionably not IL-1 or TNF. Reports by other groups shortly confirmed these findings: here was a new body message.[3]

Following a script that should be familiar by now, as research on hepatocyte-stimulating factor was proceeding, a parallel set of investigations were attempting to identify mediators active in a completely different system, B lymphocytes, the cells that manufacture antibodies. One of antibodies' primary roles is to help other components of the immune system get rid of microbes. However, to produce the right amount and type of antibodies, and thus behave at their best, B lymphocytes themselves need help from the rest of the immune system, particularly from T lymphocytes. Exactly which form or shape this help might take was not clear: Did the two kinds of lymphocytes need to touch each other, the Ts giving a hand to the Bs? Or was touching unnecessary, with molecular communication at a distance being sufficient?

Studies in the early 1970s demonstrated that a soluble mediator, a body message made by T lymphocytes, could do the trick of helping B lymphocytes without need for the Ts and Bs to get physical. Actually, T lymphocytes seemed to produce multiple such mediators, each one nudging B cells in its own slightly different way. As usual, this still rather confusing communication soup received a bunch of names: B cell-stimulating factor (which came as type 1 and type 2), B cell growth factor (also available in two flavors), and a few more. As for the discovery of IL-1 and TNF, and like scientists trying to determine the identity of hepatocyte-stimulating factor, researchers studying molecules that acted on B lymphocytes began by isolating the proteins responsible for the biological changes of interest and later went hunting for their genes.

These intense efforts culminated in 1986 with identification of the genetic sequences of three cytokines that targeted B lymphocytes and that were eventually named interleukin-4, -5, and -6. Interleukin-6 is the one we are going to follow up with, but let us first spend a few words on its immediate predecessors.

Antibodies, also called immunoglobulins (Ig in short), exist in various forms, all produced by B lymphocytes but each made under different conditions and having somewhat distinct roles. Mediators that tell B lymphocytes to change the kind of antibodies they should produce are called switch factors, as they make B lymphocytes switch from producing, say, antibodies of the IgG type to those of the IgE type. This is rather important, since it can, for example, determine the difference between having and not having allergies. Interleukin-4 is such a switch factor, crucial for production of IgE, the version of antibodies involved in the pathogenesis of allergic reactions and many other vexing situations. Interleukin-5 also partakes in these bothersome conditions, particularly in its ability to activate a type of leukocytes called eosinophils. Not surprisingly, attempts at blocking IL-4 and IL-5 for treatment of allergies and asthma are ongoing.[4]

Back to our main story. Shortly after identification of the gene coding for one of the B cell-stimulating factors (BSF-2, to be precise), parallel lines of research merged into a single track. To the surprise of many, B cell-stimulating factor-2 and hepatocyte-stimulating factor turned out to be one and the same molecule, as Jack Gauldie, Heinz Baumann, and their colleagues first demonstrated before the cytokine had even been admitted into the interleukin family with its current name of IL-6. Just as it had been for IL-1 and TNF, a single message was responsible for seemingly unrelated, disparate activities, from stimulation of antibody manufacturing by B lymphocytes to induction of acute-phase proteins by the liver. The new cytokine also coincided with a mediator that investigators had named hybridoma/plasmacytoma/myeloma growth factor due to its ability to enhance growth of myelomas and plasmacytomas, cancers of the bone marrow that derive from malignant transformation of B lymphocytes.[5]

Since the very beginning, therefore, researchers knew that IL-6 was endowed with the power to exert potentially advantageous activities, such as induction of antibodies that help the body fight infections, as well as less amiable ones, for example incitement of cancer cell proliferation. Whether the ability of IL-6 to stimulate the hepatic acute-phase response should be considered part of the friendly or the nasty side of IL-6 investigators were not sure about.

Down the IL-6 Path

If a single person should be awarded the honorific "Dr. IL-6" that would be undeniably Dr. Tadamitsu Kishimoto. Born in Tondabayashi, a small town in the southern part of Osaka Prefecture in Japan, as the only son of a high school teacher, Kishimoto was educated all the way through high school in a rural area, enjoying the natural life of the countryside. While still in primary school, he read the biography of Dr. Hideyo Noguchi, a microbiologist who made important discoveries related to syphilis and other infections, thus becoming interested in medical research and eventually graduating with an MD from Osaka University in 1964. There he met his mentor, Professor Yuichi Yamamura, who drew young Kishimoto into studying antibodies to try to understand why our immune system reacts against its own components, causing autoimmune disease.

After spending some time in the laboratory of Dr. Kimishige Ishizaka at Johns Hopkins University in Baltimore, Maryland, in the mid-1970s Kishimoto returned to Osaka University, where he has been active ever since as professor, chairman, dean, and president. It was at Osaka University that his research group, then composed of about fifty postdoctoral fellows and graduate students, in the 1980s first purified the protein and then identified the gene for IL-6, following the B lymphocyte-stimulating factor thread we just discussed. "I did not imagine that this molecule had such pluripotent activities. The discovery of IL-6 was a mix of a clear hypothesis and chance," Dr. Kishimoto told me during our interview.[6]

For many a scientist this would be enough of a discovery for a whole lifetime, but not for Dr. Kishimoto, who continued to drive his research group deeper and deeper down the IL-6 path. As Kishimoto explained, "In the 1980s, after the isolation of IL-6, within several years we isolated its receptor, the signal transducer gp130, the transcription factors NF-IL-6 and STAT-3, and the negative regulator SOCS. At that time, there were no competitors in the world. Only STAT-3 was published at the same time by Dr. [James] Darnell at the Rockefeller University and we could not identify JAK kinase before Dr. [James] Ihle. Otherwise, all the essential molecules for signal transduction from IL-6 to downstream gene expression were identified by our group. Moreover, we could find several diseases in which over-production of IL-6 was closely involved. No one could compete with us." Chapeau![7]

What are all these strange names Dr. Kishimoto mentioned? In 1988, only two years after finding the gene for IL-6, his group identified the receptor for this cytokine, the protein on the surface of cells to which IL-6 needed to tether to make the cell sense its presence and respond with a biological change. Except that—surprise, surprise!—this receptor appeared to be unable to transmit any signal inside the cell, it was way too short and lacked essential pieces. Hmmm . . . how could the message IL-6 effect a biological change with such an incompetent receptor?

Undeterred, it took Kishimoto only a couple more years to direct his team to a further critical discovery. Indeed, the group soon identified the missing piece, which was a membrane protein able to associate with the IL-6 receptor after IL-6 had attached to the receptor itself. Only at this point, with IL-6 bound to its receptor and the new protein closely associated, could a signal be activated, allowing IL-6 to deliver its message to the target cell. The researchers chose the name gp130 for this new protein to indicate its molecular size. Importantly, gp130 eventually turned out to associate with several other cytokine receptors and to be present on the surface of basically every cell in the body. Concomitant studies by other investigators that revealed the structure of several cytokine receptors and their associated proteins, as well as how these complexes transmitted signals inside cells, tremendously helped the researchers at Osaka University toward these discoveries, as Dr. Kishimoto readily acknowledged in the interview.

Now that they had the cytokine and its receptor complex, the next burning question for Kishimoto and his collaborators concerned the nature of the intracellular signals activated by IL6. In 1990, they found the first one, NF-IL6 (which stands for nuclear factor IL-6 and later became C/EBP-β), a small protein that not only granted cells responsiveness to IL-6 but also induced synthesis of IL-6 itself. Not happy with just one pathway, the researchers followed up on findings by other investigators on a molecule called acute-phase response factor (the name speaks by itself). They identified that gene, too, publishing their results simultaneously with the group of James Darnell, who named it STAT-3 (signal transducer and activator of transcription-3), the name this molecule still holds. To become active, STAT-3 needs help from another molecule, a member of the JAK (Janus-activated kinase) family: this is the sole link in the chain Dr. Kishimoto lamented not having been able to get to before others. What has become famous as the JAK-STAT pathway is now the subject of thousands of studies and a potential pharmacological target for treatment of chronic inflammation and cancer.[8]

To summarize, as soon as IL-6 binds to its receptor on the cell surface, gp130 associates with the complex. This leads to activation of a JAK protein, which in turn activates STAT-3, which then moves to the nucleus where it binds to specific sequences of DNA. As a result, genes are switched on and start making proteins that go about doing their respective jobs.

Is this system now turned on forever? Of course not. Quenching molecular signals is essential to prevent damage, particularly when it comes to dangerous pro-inflammatory cytokines. This can be achieved in the extracellular space by way of anti-inflammatory mediators, receptor antagonists or molecular decoys, as mentioned in previous chapters. Evolution, though, has also found ways to press the off button from a cell's insides: induction of suppressor of cytokine signaling (SOCS) is one of them. The SOCS family includes various members, small proteins discovered by several investigators—including Dr. Kishimoto—in the second half of the 1990s. When IL-6 and other body messages turn on the switch for activation of the JAK-STAT pathway, the response eventually ends up inducing members of the SOCS family, which then shut down the JAK-STAT system itself in a classical negative feedback loop.[9]

Now that we've followed Dr. Kishimoto all the way down the IL-6 path, from synthesis of IL-6 right into the very nucleus of the responding cell, we can gather all this information into a tryptich that provides some answers while—as any good science should do—opening up new questions.

Most cells make IL-6. Though activated leukocytes are quantitatively the most prominent source of IL-6, endothelial, muscle, and fat cells can also produce this cytokine, as can many other cell types.

Most stimuli induce synthesis of IL-6. Be it an infectious agent or another cytokine, physical or chemical stress, contraction of a muscle cell or engorgement of an adipocyte, whenever deviation from homeostasis is present you can be sure IL-6 is going to be produced. Once made, IL-6 flows outside the cell that manufactured it, acting locally as any good cytokine is supposed to behave but also swimming around in the bloodstream to reach the rest of the organism, like a hormone.

And most cells can respond to IL-6. Actually, very few cell types have IL-6 receptors on their surface, those receptors Dr. Kishimoto identified in 1988. These fortunate few are hepatocytes, B lymphocytes, and some other leukocytes. But the fact is, the majority of cells do not have receptors for IL-6 on their surface and yet are able to sense this body message. How can this be?

The Unforeseen: Trans-Signaling

Here comes the unexpected twist about soluble receptors announced at the end of the previous chapter. Rather than behaving as molecular decoys as most other cytokine soluble receptors do, the soluble receptors for IL-6 instead act in exactly the opposite way, forcibly pushing cells that don't have IL-6 receptors to hear that cytokine's message nevertheless. Isn't that a bit weird? Let's ponder the situation here. Cells that evolved and developed without the ability to respond to IL-6, choosing not to show the receptor for IL-6 on their surface, are coerced to respond anyhow by way of IL-6 soluble receptors inserting their foot into the door (the cell membrane) once they have uploaded their ligand. The other necessary piece of the triad, gp130, is already there patiently waiting—et voilà—IL-6 has now brought its message to cells that were unwilling to hear it. The presence of this mechanism, known as trans-signaling, is a peculiarity of IL-6: here is a potentially strong pro-inflammatory mediator that does not have a specific inhibitor, a cytokine that has actually managed to turn an impediment into an assistant.[10]

Although several investigators contributed to this knowledge, the scientist who devoted most of his research to the study of trans-signaling is Dr. Stefan Rose-John, currently executive director of the Institute of Biochemistry at the University of Kiel, in northern Germany. "I was an enthusiastic boy-scout and spent all of my early youth hiking and camping in the woods," Dr. Rose-John wrote in reply to my probing questions about his life. This early love of nature was one of the reasons motivating him to get a PhD in plant biochemistry. After working for a couple of years on deepwater rice in Michigan, however, Rose-John realized his interest was more on the biochemistry side than the plant side of science and decided to change focus. "By this time I had published six solid papers and was just turning thirty years old. I decided to leave the plant field and took a position as junior group leader at the German Cancer Research Center in Heidelberg," he told me. He worked there for a while, focusing on a protein called protein kinase C (PKC, of which he identified a mouse variant in 1988), but moved again rather shortly, to the Institute of Biochemistry at the University of Aachen. Professor Peter Heinrich had just been hired to direct that institute: a biochemist of liver proteins, Dr. Heinrich had done important studies on hepatocyte-stimulating factor (you should recall this as one of the early names of IL-6) and worked with Dr. Kishimoto on IL-6 soon after that cytokine's identification. Stefan Rose-John joined the quest and began investigating IL-6 himself; his very first publication on the topic, in March 1990, was a collaboration with Drs. Kishimoto and Heinrich, with whom he kept up the teamwork over the years. Soon, however, Rose-John concentrated on IL-6 receptors, following up on results obtained by some of the scientists we previously met.[11]

In 1989, the unstoppable Tadamitsu Kishimoto had shown that IL-6 bound to a manufactured soluble IL-6 receptor: the complex then associated to gp130 molecules present on the cell surface and transmitted a signal inside the target cell. This was rather unusual but also very artificial, since nobody yet knew whether soluble IL-6 receptors actually existed in nature.[12]

Concomitant work by Daniela Novick and her collaborators at the Weizmann Institute in Israel provided that needed evidence by detecting soluble IL-6 receptors in nuns' urine using affinity chromatography. In their 1989 article, Dr. Novick and colleagues described isolation of the soluble receptors for two cytokines, interferon-γ and IL-6. The article contained bioassay data demonstrating that the soluble interferon-γ receptor acted as a decoy, inhibiting that cytokine's activity just as everybody was expecting it to do. Oddly enough, even though it also reported isolation of soluble IL-6 receptors, the article made no mention of a bioassay testing for these molecules' decoy activity. "We actually did that," Daniela Novick told me. "We obviously expected the soluble receptor to inhibit IL-6 activity just as all the other known soluble receptors did. But it did not. To the contrary, IL-6 soluble receptors potentiated the activity of IL-6. At that time I thought that I did not know how to work! So did Michel Revel [Dr. Novick's past advisor and an IL-6 expert]. This is the reason why the bioassay is not included in our 1989 publication on soluble cytokine receptors. I still have those lab notebooks," which she promptly shared.[13]

Various versions of "expect the unexpected" quotes are used and abused in every field of human endeavor. But when the unexpected is not simply a knowledge twist but the exact opposite of established paradigms, it is doubt and hesitation, not a flash of insight, that rightly takes priority in the mind of the scientist. It would in fact take a few years, until 1992, before Drs. Kishimoto, Novick, and Rose-John would independently and conclusively demonstrate the anti-decoy function of soluble IL-6 receptors in cells cultured *in vitro*. Proof that this same mechanism took place *in vivo*, in the whole organism, did not surface until a 1995 article by Dr. Rose-John.[14]

By showing that soluble IL-6 receptors rendered any cell responsive to IL-6, even those without IL-6 receptors on their surface, these reports introduced a whole new way of thinking about body messages: from a specific response limited to cells that expressed receptors on their membrane to a generalized reverberation potentially involving the whole organism. It is thus all the more noticeable that the articles describing these findings did not appear in the most prestigious periodicals that usually report major discoveries but were instead published by specialized, niche journals. It is as if even the scientists who generated those results were shy in recognizing the far-flung implications of their own findings.

On top of investigating the activity of soluble IL-6 receptors, Dr. Rose-John also focused on the mechanism by which these receptors detached from the cell surface. Fusing his prior biochemical interests with research on IL-6 receptors, he demonstrated that activation of

PKC triggered production of enzymes that cut IL-6 receptors, shedding them from the surface of cells and turning the receptors into soluble versions that slosh around in body fluids and blood, eventually turning up in urine, where the Israeli scientists had found them. This was only a few years after researchers had discovered shedding of TNF receptors by specific enzymes, as discussed in the previous chapter. There was, however, some controversy about this shedding mechanism when it came to soluble IL-6 receptors, since other scientists had proposed an alternative explanation for their becoming soluble, an explanation that, as Dr. Rose-John told me, "was viewed by the scientific community as being much more likely than ours. Therefore, we had a hard time to publish our findings in high-impact journals. The shedding of the IL-6 receptors and the pathway of IL-6 trans-signaling were very enigmatic in the beginning. But I felt that this was important, although everyone else thought it was not. Obviously, you should listen to what other scientists tell you, but you may also follow your story and make your point. For this you need a lot of strength and persistence, but it might pay off in the end. Only, you do not know this in the beginning." Following his instincts, Rose-John painstakingly continued to deconstruct the IL-6 trans-signaling pathway. Accumulating evidence eventually ended up showing that enzyme-induced shedding was likely the most prominent mechanism for generating soluble IL-6 receptors. Regardless of how IL-6 receptors become soluble, however, their presence nevertheless imparts IL-6 responsiveness to virtually every cell in the body by activation of trans-signaling, a term coined by Drs. Rose-John and Heinrich in a highly cited 1994 review article.[15]

A Peculiar Cytokine

Dangerous and protective, pro-inflammatory and anti-inflammatory, multifunctional, complex: these and many other adjectives have been used to describe the biological activities of IL-6, which has been called a sheep, a wolf, and a wolf in sheep's clothing. It is quite remarkable how many things this little protein is and does: teasing apart when and how it behaves one way or another has not been easy. If IL-1 is multifunctional in its ability to stimulate numerous responses, at least this cytokine carries an unambiguous message, that of promoting inflammation. Not so for IL-6, for which a more appropriate definition should include the terms ambivalent and contradictory.

As mentioned, many stimuli induce synthesis of IL-6, which can be made by almost any type of cell; moreover, essentially all cells respond to IL-6, albeit via distinct mechanisms. By being everywhere and touching everything, this cytokine is therefore bound to be both kind and nasty, indeed impersonating both the sheep and the wolf.

Some of the kind sides of IL-6 include helping the immune system fight infections, assisting in closing wounds and repairing tissues, supporting the generation of leukocytes and platelets, contributing to the health-boosting effects of exercise, and curbing inflammation. On the nasty side we can list, for example, ravaging of bone and cartilage, propelling multiplication of cancer cells, hustling cardiovascular disease, precipitating fever, and fomenting inflammation.

Most of IL-6's nastiness can be interpreted as the deranged action of a molecule that for some reason has lost control over its own affairs. If we wanted to target IL-6 to treat any sort of illness, it would thus be a good idea to figure out what makes it lose control and turn from good to bad. This way, one could try to obstruct it when it is nasty and support it when it is friendly.

Here are some possibilities as to what can push IL-6 to do an about-face. Up-and-down levels as compared to constant elevation may be a distinguishing feature. Thus, the seesawing blood and tissue concentrations of IL-6 caused by contracting muscles during bouts of exercise can help promote a healthy metabolism, whereas IL-6 chronically overproduced for years by the adipose tissue of obese individuals likely facilitates development of cancer. Another differentiating mechanism may have to do with how much IL-6 is made in response to a stimulus, with modest amounts increasing defenses against infections while exaggerated levels would provoke overwhelming inflammation, as in sepsis. Moreover, the context in which IL-6 is produced, that is, which other molecules are present in its immediate surroundings and thus share a conversation with this cytokine, is decisive. For example, IL-6 works in close association with another cytokine, transforming growth factor β, to generate a type of T lymphocyte that is strongly inflammatory.[16] But when the company is of another sort IL-6 switches opinion and instead increases output of anti-inflammatory cytokines while reducing generation of pro-inflammatory ones.[17] Wouldn't it be nice to have a tool to separate beneficial IL-6 from harmful IL-6?

Antibodies that block binding of IL-6 to its receptor are used to treat rheumatoid arthritis (and other forms of arthritis), the same condition physicians handle with drugs that block TNF or IL-1. Dr. Kishimoto's group initially generated these anti-IL-6 antibodies, and pharmaceutical companies took them to the clinic. "Our discovery of IL-6 and characterization of its biology led to development of drugs for treatment of many serious diseases. I think this is a good example of how a truly basic study can benefit humankind," Dr. Kishimoto told me. Other compounds that block either IL-6 or its receptor are under development by various companies for treatment of several disorders. However, as with drugs that block IL-1 and TNF, interfering with IL-6 activity causes side effects, which include increased risk of infections as well as alterations in leukocyte counts, blood pressure, and cholesterol levels. By blocking IL-6 with the current drugs, which inhibit signaling mediated by both surface receptors (termed classical signaling) and soluble receptors (trans-signaling), we may be driving away the sheep together with the wolves.[18]

However, the peculiar signaling mechanisms of IL-6 may be exploited to separate the two. Research by Dr. Rose-John and by other investigators points to differences between classical signaling and trans-signaling as one of the keys that may differentiate the hard-to-tease-apart attributes of IL-6. The idea is that trans-signaling by soluble receptors would represent the nasty sides of IL-6 you would want to block to provide relief to patients with rheumatoid arthritis and other conditions, whereas it would be better to leave classical signaling alone to provide beneficial reactions, such as protection from infections. Therefore, depending on which of the two mechanisms a cell employs to respond to the same IL-6 molecule—classical or trans-signaling—that cell's interpretation of the IL-6 message would be rather different, even opposite. Perhaps a drug that could selectively block trans-signaling, such as the one Stefan Rose-John and his team are currently testing, could disentangle IL-6 the kind from IL-6 the nasty.[19]

The Acute-Phase Response: Take Two

Toward the beginning of this chapter we asked how infection and inflammation induce production of acute-phase proteins by the liver. We now recognize this response as one of the main activities of IL-6, the origin of its early designation as hepatocyte-stimulating factor. Other cytokines and mediators, including IL-1 and TNF, can also perform this job, but IL-6 is by far the most potent. Not that IL-6 is always necessary: subtraction studies using knockout mice and other approaches clearly demonstrated that the nature of the inflammatory stimulus is decisive in determining whether IL-6 is required for induction of acute-phase proteins.[20] After reviewing IL-6 signaling modalities, we now also appreciate that hepatocytes are among the few cell types that express IL-6 receptors on their surface, making these cells responsive to classical IL-6 signaling.

We are thus left with the issue of why the liver responds to infection and inflammation with such a massive alteration in the type and amount of proteins it produces. The most common explanation for this phenomenon is that acute-phase proteins are an important component of the defense response, trapping bugs and facilitating their phagocytosis, contributing to activation of leukocytes, and lowering circulating concentrations of substances that may help microbial growth. That the acute-phase response has been relatively highly conserved throughout evolution indicates it likely plays a crucial role. However, actual proof demonstrating whether the acute-phase response truly is essential in fighting infections has been slow to come by, for reasons we are about to discuss.[21]

On top of the massive changes observed during acute illnesses, in the last three decades researchers also recognized the presence of more modest, chronic elevations of acute-phase proteins in people affected by most of the ailments of modern societies, chief of all cardiovascular disease but also obesity, diabetes, autoimmunity, and many more. This observation is not that surprising, since inflammation—which increases synthesis of acute-phase proteins—is at the root of all these disorders. Clinicians can in fact use levels of acute-phase proteins, most commonly CRP, as sensitive markers to gauge the presence of ongoing inflammation. But there is still uncertainty as to whether we should praise or blame acute-phase proteins, whether we should try to preserve (and maybe encourage) their production as a helpful reaction to the presence of chronic inflammation or we should instead find ways to prevent or block them because they contribute to poor health. Taking a look at a few acute-phase proteins may help us understand why there are still so many unknowns in this area.

C-Reactive Protein

Discovered in the 1930s for its ability to attach to pneumococci, CRP is a pretty protein that looks like a tiny flower. It belongs to a class of proteins named pentraxins for their ability to form such flowery structures when five (penta) subunits of the protein assemble into a ring.[22]

We know a lot about what CRP can do: it can bind to bacteria (of various kinds, not simply the pneumococci that sparked its discovery and helped name it) and interfere with their growth; it can activate various components of the immune response, enhancing the body's defenses; it can modulate production of cytokines and other mediators, resulting in both pro- and anti-inflammatory effects. It can have multiple additional activities that sometimes appear paradoxical. However, while we have many answers to the question, "What can CRP do?" we do not clearly understand whether CRP is actually required for any of these effects inside the organism and, if it is, whether the outcome of its involvement is to be considered positive or negative in any given context. Part of the reason for this ignorance can be traced to differences between humans and mice.

Debate about CRP's biological role has been raging since this protein's discovery, when scientists first reported a CRP surge during the acute phase of infection and inflammation. This impressive reaction, which can raise CRP up to ten thousandfold, occurs in humans and in many other species, but unfortunately not in mice, the favorite experimental animals of biomedical researchers, in which CRP only rises a meager three- to fivefold during the acute-phase response. On top of this quirkiness, CRP made by humans and CRP made by mice have disparate properties, so it's difficult to directly compare the two.[23]

To get around these issues, scientists devised clever experimental approaches. Since CRP does not naturally rise that much in infected or inflamed mice, a researcher could induce an artificial surge and evaluate whether this intervention brings about detectable biological changes. And because biomedical investigators are mostly interested in the activity of human—not mouse—CRP, they can inject the human protein into mice or generate transgenic animals that make lots of human CRP. These techniques have been helpful in demonstrating that high levels of human CRP protect mice against various infectious agents: bacteria tend to grow less well in mice with artificially raised CRP, while inflammation is tweaked, ultimately resulting in CRP-enhanced animals surviving infections better than mice in which CRP had not been pumped up. Studies looking at the role of CRP in experimental models of autoimmunity have come to similar conclusions: more CRP, less severe disease. We therefore seem to have a good player here, with data supporting the notion that the high CRP induced by infection and inflammation is likely beneficial, helping to fight off pathogens while also keeping autoimmunity under control.[24]

However, we easily recognize these experiments as addition approaches that tell us whether CRP can mediate a certain biological effect but do not say anything as to whether this protein's presence is actually necessary for that effect to take place. We would need subtraction methods to answer this other question definitively, but this has been difficult in the field of CRP. Although scientists have known for decades how to generate knockout mice—the ultimate *in vivo* subtraction methodology—it wasn't until 2011 that researchers genetically engineered mice without the CRP gene. This tardiness had nothing to do with technical difficulties but should rather be blamed on the oddity of CRP production in mice, which led scientists to discount these animals as useful models for subtraction studies. Because CRP never gets very high in mice, the reasoning was that removing its gene would not make much experimental sense. The few studies that eventually looked at the effect of CRP deletion in mice generally supported the findings of the addition approaches mentioned above: mice that did not produce CRP were more susceptible to certain bacterial infections (though not all) and got worse autoimmune disorders.[25]

Debate, however, continues, and rightly so. Because mice do not mount a strong CRP response, these subtraction studies, powerful as they are, fail to answer the question about the necessity of the CRP surge during the acute-phase response. If scientists wish to interrogate directly the need for the high CRP levels observed in humans during the acute phase of disease, subtraction studies in animals that, like humans, respond to infection and inflammation with a sharp CRP rise—rabbits for example—would be needed.[26]

Nevertheless, even if CRP subtraction studies in mice cannot conclusively address the role of the CRP surge, what they do reveal is the involvement of the modest, basal levels of CRP that are present in blood on a daily basis, an issue that brings us directly to one of the most hotly disputed topics in the field of CRP: the involvement of this protein in the pathogenesis of cardiovascular disease.

Thousands of studies, collectively examining hundreds of thousands of individuals, conclusively concluded that high CRP predicts risk of cardiovascular events such as heart attacks. We are not talking here about the sky-high CRP levels seen in response to acute infections, but about basal CRP, those moderate concentrations present in the bloodstream outside of bouts of illness. People who have raised basal CRP levels are more likely to have coronary heart problems than those with basal CRP toward the low end of the spectrum (all other variables being equal).[27]

Elevated daily CRP is generally (albeit not always) caused by underlying smoldering inflammation that induces a limited, though significant, increase in IL-6 and other cytokines. Scientists have wondered for a long time whether CRP is simply a bystander, a marker of the underlying inflammation, or instead participates in the pathogenesis of cardiovascular disease. Cardiovascular disease being the chief killer in the world, the question is obviously significant.

Experimental studies mostly performed using the addition approach have in large part fueled controversy over the possible role of CRP in cardiovascular disease. Worsening, amelioration, lack of effect: animal studies have demonstrated everything and its opposite, with fierce arguments, almost down to the personal, as to which experimental models are appropriate to answer this specific question and which are not. *In vitro* experiments, though helpful, have also failed to generate definitive answers.[28]

Human diversity is what helped scientists disentangle these contentious knots. Here is how. The CRP gene comes in multiple variants, some of which are associated with production of more CRP than average regardless of the presence of underlying inflammation. Studying the risk of cardiovascular disease in relation to this genetic variability can help researchers understand whether CRP is cause or effect. We know that, in general, people who have elevated basal CRP are at increased risk of heart attacks. What we are asking now is whether those individuals whose CRP is elevated simply due to genetic variability are also at increased risk of heart attacks. In other words, we are trying to find out whether a high basal CRP by itself—in the absence of inflammation—is sufficient to augment the risk of cardiovascular disease, which would reveal a pathogenic role for CRP that is independent of that of inflammation.

For once here we have a clear answer. Studies started in the mid-2000s, and by now including tens of thousands of individuals, demonstrated that CRP that is elevated due to genetic variability is not linked to cardiovascular disease risk. It is only when basal CRP rises because of underlying inflammation that the risk rises. What this extensive evidence tells us is that CRP is the marker rather than the maker in cardiovascular disease, a sign that an inflammatory response is ongoing but not a significant player in determining the presence or absence of disease by itself.[29]

Even though genetic diversity indicates that CRP is not part of the pathogenetic pathway of cardiovascular disease, additional studies indicated that certain variants of the IL-6 receptor are instead associated with reduced risk of cardiovascular-related disorders. These data clearly point to IL-6, and therefore to the inflammatory response, as part of the mechanism of cardiovascular disease and cycle us back to our peculiar body message and its plentiful biological activities.[30]

When we put all the evidence we just reviewed (and much more) together, we can conclude that chronic, simmering inflammation caused by unhealthy lifestyles that disrupt metabolic pathways and perturb blood vessels is the main cause of cardiovascular disease. This chronic inflammation induces IL-6, which contributes to pathogenesis (as we can deduce from genetic variability studies examining variants of the IL-6 receptor) while also raising basal levels of CRP. However, CRP is unlikely to have a major pathogenetic role by itself. The story is probably different when considering the CRP surge of acute illness, but evidence in that case is not yet completely convincing.

If the biology of CRP made you woozy, you won't feel much steadier as we proceed to the next acute-phase protein.

Serum Amyloid A

Sea cucumbers, carp, duck, minks, hamsters, mice, rabbits, dogs, humans (and many more species): when any of these animals gets infected with a bug, production of serum amyloid A (SAA) shoots up. What we call SAA is actually (in humans) three proteins coming from three separate genes, SAA-1, SAA-2, and SAA-4 (SAA-3 is a pseudogene, i.e., a gene that does not encode a protein). Like CRP, SAA-1 and SAA-2 are acute-phase proteins, produced in massive amounts by the liver in response to infection and inflammation: because they are so similar to each other, I will use SAA to refer to both of them. Their cousin SAA-4 prefers to set itself aside, being made by several types of cells and not responding to inflammation: I'll leave it alone.

"The exact function of SAA is not too clear." These are the words of Dr. Jean Sipe, who spent the major part of her scientific life studying SAA. Now retired after a long career, Jean Sipe was born and raised in Decorah, Iowa, in what she described as a nurturing community of mainly Norwegian immigrants. Education was highly valued in her family, and in the community at large. She was always a strong student in mathematics and science. When the Soviet Union launched the Sputnik spacecraft in 1957, Jean's teachers encouraged anyone talented in these areas to go into science as the patriotic thing to do. She accepted the challenge and enrolled at Iowa State University to study chemistry, which she had enjoyed in high school. "There I received an excellent education," she told me. She was then off to the University of Washington in Seattle to experience the Pacific Ocean while earning a master's of science in biochemistry, and later got her PhD in chemistry focusing on lipid metabolism at the University of Maryland, after having worked a bit in a virology laboratory. In the 1970s, Dr. Sipe was offered a staff position at the NIH in the laboratory of Dr. George Glenner, whose pioneering studies of amyloidosis led her directly to SAA.[31]

The term *amyloidosis* refers to a group of disorders characterized by deposition of large amounts of proteins in tissues, where they form structures that resemble small fibers, or fibrils, visible under the microscope. Many proteins can do this (we know of about thirty by now) and there are many reasons why this may happen, from genetic mutations to chronic infections to the yet unknown. Kidneys, heart, and the nervous system are most commonly affected in amyloidosis, with deposition of fibrils altering the structure of tissues and organs, eventually leading to loss of function. Alzheimer's disease is probably the form of amyloidosis most people are familiar with: here protein fibrils accumulate in areas of the brain involved in cognition, causing progressive memory loss and other alterations.[32]

The version of amyloidosis Dr. Sipe was studying in the 1970s and 1980s was not Alzheimer's disease but a form called systemic amyloidosis, where fibrils wind up accumulating in more than one organ, not just the brain. Systemic amyloidosis itself comes in various flavors, one of which is designated as reactive, or secondary, because it develops as a secondary reaction to other conditions, typically chronic infectious or inflammatory disorders.

Rudolf Virchow, the German physician-scientist regarded as the father of modern pathology, first described systemic amyloidosis in the late 1800s. In nineteenth-century Europe, tuberculosis was the most common condition to which the organism reacted to with amyloidosis; it shamefully still is in those parts of the world where this chronic infection remains a scourge. In the richer parts of today's world, however, where tuberculosis is infrequent, reactive amyloidosis mostly originates from the chronic inflammation of disorders like rheumatoid arthritis and inflammatory bowel disease. Effective therapies for these conditions, described in this and previous chapters, have helped bring down even these forms, so that reactive amyloidosis itself has become fairly rare in the richest parts of the globe. You must have guessed it by now: SAA is the major protein that forms the fibrils that accumulate in tissues of people with reactive amyloidosis (there are others, but SAA is the most common).[33]

We have pretty good knowledge of what regulates synthesis of SAA by the liver and, in large part, can thank Dr. Sipe for that. After making important contributions to the biochemical characterization of SAA, Dr. Sipe was in fact among the first scientists to study the inflammatory mediators that upregulate production of this acute-phase protein in hepatocytes, often collaborating with some of the investigators we previously met to demonstrate the involvement of IL-6 and other cyto-kines in this process.[34]

Besides regulation of its output by the liver, however, there are few other certainties about the biology of SAA. The strong degree of conservation of its gene throughout evolution indicates SAA must have important roles, but, as Dr. Sipe mentioned and other scientists con-firmed, we do not clearly understand what these might be. Most of the SAA that circulates in blood is attached to high-density lipoproteins (HDL) that also carry good cholesterol, the form of cholesterol that pro-tects against cardiovascular disease. When SAA surges in response to infection or inflammation there is so much SAA attached to HDL that it ends up modifying HDL's properties and reducing its protective actions. Furthermore, SAA can activate a whole range of receptors, resulting mostly, but not exclusively, in promotion of pro-inflammatory pathways. Interestingly, SAA shares a receptor with Anxa1 and lipoxins, substances that, as presented in Chapter 4, mediate the resolution of inflammation. Remarkably, though, binding of SAA to this receptor elicits pro-inflammatory outcomes, whereas the opposite happens when Anxa1 and lipoxins attach to it. How this comes to be is still unknown.[35]

These and several other activities implicate SAA in the pathogen-esis of the same conditions in which CRP has also been involved: car-diovascular and autoimmune disorders, obesity and diabetes, infections and more. But whether SAA is required (or even simply participates) in causing, perpetuating, or protecting from any disease other than reac-tive amyloidosis remains to be elucidated.[36]

Production by the liver, induction by infection and inflammation, participation in the transport of cholesterol: SAA straddles the border between metabolism and inflammation. The next acute-phase protein we are going to discuss—hepcidin—is of even more difficult classifica-tion even though, for a change, its function is clear.

Hepcidin

Hep for hepatic, because it's the liver that makes it; *cidin* for microbi-cidal protein, because it was first discovered as a killer of microbes. Identified through such a series of fortuitous events that make it stand out among all body messages, hepcidin is the body's main regulator of iron metabolism. For hepcidin, structure came first, accompanied by what is likely a minor biological function for this protein, while its crucial role as a determinant of iron distribution in the body was found by mistake.

In 1998, while isolating antimicrobial compounds from human urine, Dr. Tomas Ganz and his research group at the University of California, Los Angeles, found a small protein with a distinctive structure: it came from the liver and was able to slow the growth of some fungi and bacteria (although it wasn't very good at that), so the researchers named it hepcidin. As it happens, another set of investigators, directed by Dr. Knut Adermann in Hanover, Germany, independently found the same molecule at about the same time; they gave it a different name that nevertheless conveyed the very same meaning, liver-expressed antimicrobial peptide-1. Meanwhile, as they were pursuing a completely separate goal, Olivier Loréal and his collaborators in Rennes, France, discovered that iron could regulate expression of hepcidin in the liver. Dr. Loréal's studies and those of Dr. Ganz also indicated a link with inflammation, which sharply increased hepcidin's levels, an issue to which I shall soon return.[37]

Eventually, a technical quirk revealed hepcidin's true function. That this happened simultaneously with the discoveries we just mentioned is simply a marvel of coincidence. Sophie Vaulont's team in Paris, France, had genetically engineered mice to lack a gene, called USF2, that is involved in glucose metabolism. What Dr. Vaulont found, though, was something utterly unexpected: iron levels were all messed up in these knockout mice, even though the gene the investigators had deleted apparently had no connection whatsoever with iron. Careful analysis revealed that the genetic manipulations performed to generate USF2 knockout mice had unintentionally interfered with the hepcidin gene—which sits adjacent to the gene coding for USF2—completely shutting down synthesis of hepcidin. The Parisian researchers had therefore involuntarily generated a functional hepcidin knockout mouse, accidentally revealing hepcidin's main activity, that is, that of being iron's master switch.[38]

To appreciate fully the importance of this discovery we should take a detour into the world of iron. But first allow me a note about the Three Princes of Serendip, who, during their travels, "were always making discoveries, by accidents and sagacity, of things which they were not in quest of." It is from this (probably) Persian tale that in 1757 Horace Walpole coined the term *serendipity*, a word currently defined as "the faculty of making fortunate discoveries by accident" (and variations thereof). I find it unfortunate that the sagacity piece of the old tale completely dropped out of the modern definition of serendipity. Otherwise, serendipitous would be the perfect description for the discovery of hepcidin and of so many other scientific breakthroughs that happened when scientists looking for one thing were paying enough attention to the whole of their experimental system to detect unexpected outcomes that turned out to be more consequential than the original plans. Accidents without sagacity do not lead very far, as the Three Princes certainly knew.[39]

Iron, then. This metal is essential for survival: it is to iron, complexed to heme as a component of hemoglobin in red blood cells, that oxygen attaches to so it can be carried from the lungs to each little corner of our organism. Not enough iron, not enough oxygen to cells. Iron deficiency is the most common nutritional disorder worldwide, affecting mostly women and children, as well as the most common cause of anemia, a reduced ability of the body to carry oxygen. On top of transporting oxygen, iron is also required for several biochemical reactions that keep cells functioning properly. Though our body needs an appropriate supply of iron, having too much is undeniably toxic. A genetic disease called hereditary hemochromatosis leads to accumulation of iron in several organs, particularly the liver, where this metal wreaks havoc, damaging cells and promoting cancer. Even without reaching these extremes, it is better not to have extra iron on top of what our organism needs.

Investigators have been studying iron metabolism for ages with these very issues in mind: something that can turn from a substance that nurtures life to an element that hinders it. Since iron can be dangerous, evolution must have come up with ways to fine-tune its levels. Interestingly enough, however, we cannot actively excrete iron, probably because iron deficiency has been definitely more common than iron excess throughout our evolutionary history, so it made sense for the body to save as much iron as possible. If we cannot modulate iron excretion, we have to deal with whatever amount enters our organism, meaning we should have ways to regulate the quantity of iron we absorb from the food we eat. When our organism needs a lot of iron—during pregnancy, in a growing child, or after major blood loss, for instance—it makes perfect sense to be able to extract all the iron we can from our food. Instead, it would seem smarter to shut down iron absorption once stores have been replenished, so we do not end up with toxic concentrations. Scientists had known since the 1960s that iron is absorbed by cells of the small intestine and speculated that a humoral factor, some sort of body message, would regulate its absorption. It took four decades but the Three Princes of Serendip finally found hepcidin.[40]

Let's see how the system works. Enterocytes, the cells that line the intestine, have a side that looks toward the intestinal lumen, where food arrives from the stomach, and another side that connects with the circulatory system, so they can transfer nutrients from food to blood. Enterocytes pick up as much iron as they can from food using proteins that are present on the surface that looks to the lumen. Once iron is inside enterocytes, it moves to the other side to get into the blood so it can be distributed where it is needed. To exit from the enterocyte and go into the blood, iron needs to pass through a protein called ferroportin, the iron door. When ferroportin is present the door is open, iron can reach the bloodstream and thus be properly absorbed and distributed. However, if there is no ferroportin the door remains closed and iron is trapped inside enterocytes, unable to reach the blood.[41] It is the job of hepcidin to decide whether the iron door stays open or closed. Hepcidin attaches to ferroportin and destroys it, locking the door until enterocytes synthesize new ferroportin. When individuals are in a state of iron deficiency—anemic because they lack iron or have had a hemorrhage for example—the liver shuts down production of hepcidin, so there is plenty of ferroportin to let in all the iron that's available. When iron stores are replenished, hepcidin bounces back and the door stays only partially open; if iron rises above a critical threshold, hepcidin's output goes up even further and the iron door is sealed. Therefore, hepcidin is a negative regulator of iron absorption: the more hepcidin there is, the less iron can go from food to blood.

The scientist who figured this out did not come from Serendip; she nevertheless had to travel halfway around the globe to reach Dr. Ganz's laboratory in Los Angeles. I will let Dr. Elizabeta Nemeth, now herself a professor at the University of California, Los Angeles, tell her own story.

I was born in Yugoslavia in 1969, in a university town called Novi Sad. I had an absolutely wonderful childhood, loving parents and grandparents, a lot of friends to play with in the building where I lived, and a warm and supportive community. I loved to read books from an early age, and my parents fostered curiosity: some of my favorite memories were those of nature expeditions.

I was interested in science and lucky to cross paths with wonderful, dedicated mentors who provided rich opportunities to fall in love with it. In elementary school, I was very focused on math (I did math as an extracurricular activity and went to math competitions), but then molecular biology captured my interest in high school. I was in a math and sciences high school and we had a requirement to do a high school thesis, that is, conduct and write up a research project. I did that at the prenatal diagnostic center of the medical school: they were just setting up protocols for chorionic villus sampling and let me participate in the project. They taught me how to isolate chromosomes and identify different chromosomal aberrations, and I was hooked.

After high school, I attended the program of Molecular Biology and Physiology at the University of Belgrade. As part of the program, we had to complete an undergraduate thesis by working in a research lab. However, civil war in Yugoslavia broke out during my undergraduate studies and it became extremely difficult to do scientific research. I had the good fortune of joining the lab of a wonderful and driven neurobiologist, Dr. Sabera Ruzdijic, who kept the lab going during extremely difficult times through rare dedication and ingenuity.

I graduated from the University of Belgrade as the country was amidst civil war, going through an economic and societal collapse. Like many other college graduates from Yugoslav universities, I decided to leave the country and attend graduate school elsewhere. Even the process of applying was extremely difficult. The country was under strict international embargo and we could not take TOEFL or GRE tests in Yugoslavia, we could not mail applications to international universities or pay application fees by bank transfer. I did not know anything about American universities but, fortunately, the United States consulate in Belgrade had a library that carried the Peterson's guide to graduate studies. That was instrumental in my ability to choose universities and figure out how to apply—it was 1993, no Internet available yet.

In February 1994, I received a phone call from the University of Hawaii in Honolulu offering me a position in their interdisciplinary graduate program in cell, molecular, and neurosciences. I took it on the spot; I was so happy and grateful to feel that my future might not be bleak after all. At the University of Hawaii, I did my PhD studies with Dr. Gillian Bryant-Greenwood on the mechanisms of preterm birth. This was my introduction to American science and I decided that, if possible, I was going to do academic research as a career.

After my PhD and a short postdoc at Cedars-Sinai on ovarian cancer, I was looking for a job and a friend of mine told me about a flyer in a corridor at the University of California at Los Angeles that said, "Got iron?" This was from the laboratory of Dr. Tomas Ganz. I did not know anything about iron metabolism, but was interested in learning about a new field. I applied for the position, was hired, and started working on hepcidin and iron metabolism, on which I have since focused my research. There were only four papers published on hepcidin when I started working on this topic in the early 2000s.

There are now almost three thousand scientific reports on hepcidin and a good proportion of them carry Dr. Nemeth's name. While a post-doctoral fellow with Tomas Ganz and collaborating with the group of Dr. Jerry Kaplan at the University of Utah, Dr. Nemeth was the first author of a 2004 report describing how hepcidin works by degrading ferroportin, thus locking the iron door.[42]

Even before arriving at these crucial findings, Drs. Nemeth and Ganz had conducted studies that linked hepcidin to our main topic, IL-6 and the acute-phase response. The anemia of inflammation is a common consequence of an assortment of chronic disorders that have either an infectious or an inflammatory component, from malaria to rheumatoid arthritis, from AIDS to cancer. This is the second most common form of anemia worldwide after iron deficiency. Scientists already knew, since the 1930s, that iron levels tumbled in the blood of individuals and experimental animals affected by infections or inflammation, resulting in anemia, but exactly how this happened was unknown, though involvement of the usual suspects—inflammatory mediators—had been proposed.[43]

Following up on earlier observations that hepcidin levels were elevated under inflammatory conditions, Drs. Nemeth and Ganz demonstrated that hepcidin behaved like a typical acute-phase protein, rising when hepatocytes were stimulated with IL-6 (and IL-1, TNF, and other inflammatory substances). The investigators also established that induction of hepcidin by IL-6 was often responsible for the anemia of inflammation.[44] Thus, in the context of infection and inflammation, hepcidin is similar to CRP and SAA in its ability to rise quickly in response to IL6. Unlike CRP and SAA, though, whose precise function remains a riddle, we know the exact aftermath of hepcidin's rise: the iron door is sealed, iron drops and, with time, anemia creeps in. Things are definitely clearer here. And yet, if we keep asking the question of why this happens, of what is the advantage of lowering iron levels during infection and inflammation, we still come up with an incomplete answer.

The evolutionary explanation for the anemia of infection has been that reducing iron concentrations slows down replication of microbes while boosting immunity: definitely a helpful reaction. Iron can also promote cancer growth, so keeping it low may represent a favorable outcome in patients with malignancies. However, clinicians have traditionally reacted to the anemia of chronic diseases by attempting to resolve it, supported by data indicating that patients with anemia have reduced quality of life and increased mortality. Whether physicians should indeed always attempt to redress the anemia of inflammation remains an unsettled issue, with therapies for anemia possibly reducing inflammation and ameliorating symptoms, but also potentially causing unwanted side effects. In many cases, whether low iron and anemia do or do not provide some sort of advantage to patients affected by diseases with an inflammatory component simply remains unknown.[45]

Discovered as a urinary antimicrobial compound, acting as the master hormone for iron metabolism, yet behaving as an inflammation-responsive hepatic acute-phase protein—confining hepcidin to a single box is truly an impossible task.

From Liver to Fat

Since we have gradually come to incorporate metabolism into our discussion, it seems appropriate here to announce the next chapter, where we will encounter two body messages—leptin and adiponectin—discovered in the context of metabolic research.

As much as the liver has been the protagonist of the current episode, fat (properly named adipose tissue) will be the main character of the upcoming one. At the beginning of this chapter, I mentioned the complexity of the liver's job: although mostly studied as part of metabolic research, scientists have known and intensely investigated the liver's participation in inflammatory responses for at least a century. Adipose tissue has had a similar fate, though more recent. When rates of obesity began to soar in Westernized countries in the 1990s, and later all over the world, researchers came to recognize the presence of chronic low-grade inflammation in obese individuals, mostly stemming from the expanded adipose tissue. Studies reporting elevated levels of CRP and its main inducer IL-6 in obese compared to lean individuals were among the first to connect obesity to inflammation.[46]

There are many similarities between the acute-phase response we discussed in this chapter and the chronic low-grade inflammation of obesity. Many of the same questions have been asked, some have been answered. When adipose tissue expands, particularly in the belly area, its cells, the adipocytes, engorge so much fat that they grow too big, eventually choking themselves. These giant adipocytes send out SOS messages that leukocytes acknowledge by sweeping in, probably to try to mop up the damage, as these cells always do. In the process, more immune cells get activated and inflammation ensues, with release of most of the body messages we previously met, including IL-6, which flows to the liver and induces CRP, SAA, and hepcidin, setting in motion the whole acute-phase response. This is not the major surge that occurs after an acute infection, but rather looks like the modest but chronic increase seen in individuals at risk for heart attacks. Indeed, obesity does increase the likelihood of developing cardiovascular disease and many other disorders, including some types of cancer and, especially, diabetes. Investigators have often blamed the chronic inflammation aroused by distressed adipocytes for obesity's comorbidities, those conditions whose prevalence is higher in the obese compared to the non-obese.[47]

Since IL-6 is elevated in obese individuals, is there a specific contribution for this cytokine in modulating the risk of obesity's comorbidities and, given this cytokine's biological ambivalence, would this be a protective or a deleterious effect? A few pages back we discussed evidence derived from genetic diversity studies indicating that individuals with a specific variant of the IL-6 receptor gene are at low risk of coronary heart problems. These studies and additional findings pointed to IL-6 as an important mediator in the pathogenesis of cardiovascular disease. Similar results have emerged for cancer, with hundreds of studies showing that chronic activation of the JAK/STAT pathway by IL-6 enhances tumor growth. Data therefore demonstrate that the chronically elevated IL-6 present in obese individuals likely contributes to amplifying the risk of two of obesity's frequent companions, cardiovascular disease and cancer.[48]

However, the role of IL-6 has been more difficult to define in reference to diabetes, the major comorbidity of obesity. Diabetes is an inability of the body's cells to utilize glucose (sugar) as a form of energy. Why is it bad to have diabetes? Because this condition strongly increases the risk of having serious problems with various parts of the body, including heart, kidneys, and nervous system, thus reducing both quality and quantity of life.

The diabetes of the obese is called type II, whereas type I diabetes is an autoimmune condition that mostly begins during childhood or adolescence. In type I diabetes the cells that produce the hormone insulin are destroyed by the immune system, whereas in type II diabetes the organism fails to properly respond to insulin's action and eventually insulin production itself ceases. The World Health Organization tells us that type II is by far the most common form of diabetes in the world, representing 90 percent of all cases, affecting almost 10 percent of adults worldwide and becoming ever more common in children.[49]

Most individuals with type II diabetes are obese and sedentary; the vast majority of cases can be prevented, and at least partially treated, by adhering to a healthy lifestyle. Because obesity is a major risk factor for type II diabetes, and because IL-6 is elevated in the obese, many investigators have attempted to understand whether IL-6 is a sheep or a wolf in obese individuals when it comes to regulation of glucose metabolism. In other words, does the high IL-6 of obesity represent an attempt to try to keep sugar levels in check or does it instead contribute to development of diabetes? After years of data seemingly pointing the finger to IL-6 as a diabetes villain, more recent investigations indicated that this cytokine may instead act as an important protective mediator against disruption of sugar metabolism, and that the signaling mechanisms involved (classical versus trans-signaling) may make a difference.[50]

Therefore, obesity presents a riddle, with chronically elevated IL-6 promoting cardiovascular disease and cancer but also seemingly protecting from diabetes. That diabetes itself, independently of obesity, amplifies rates of cardiovascular disease and cancer, possibly by magnifying inflammation, further complicates interpretation and understanding of these findings. These intricacies are not restricted to IL-6. For example, researchers know that an iron-rich diet favors diabetes but also that obesity, which often leads to diabetes, disrupts iron metabolism, partly due to increased synthesis of hepcidin by the liver in response to inflammation that derives from enlarged adipose tissue. Do the reduced iron levels seen in obesity represent a protective response against diabetes? And if so, is it a good idea for physicians to try to fix them?[51]

It looks like obesity throws the organism off balance, making it hard to respond with adjustments that, having evolved over millions of years in which barely anybody had to deal with having too much fat, are now clashing with an unanticipated reality.

A Modern Humoralism

As pointed out at the beginning of this chapter, the humoral theory of Western medical tradition considered sickness, of both body and mind, as an imbalance among four humors, whose existence, however, was never demonstrated. Illness did not necessarily take place in a physical space—an organ or a tissue—but was rather a distributed entity that should be approached with attempts to rebalance those humors. The anatomical theory that followed humoralism drastically modified this perspective, pointing to specific places in which specific disorders occurred. A myocardial infarction was a problem of the heart, a stroke of the brain, eczema of the skin. The germ theory, likewise, indicated specific microbes as the cause of distinct infectious diseases. We owe much of medicine's ability to reduce human suffering to this change in perspective, which allowed physicians and scientists to focus on microbial pathogens and on alterations in molecules, cells, tissues, and organs as causative of disease, leading to development of effective strategies to combat countless dreadful pathologies.

But interpretation of illness as loss of balance, as a distributed entity that does not necessarily occupy a specific space and instead results from a diffuse disruption of homeostasis, probably needs a comeback, this time supported by the scientific method (I am certainly not the first one to propose such concept). Where, in fact, should we point the accusing finger when trying to discern the physical space and the ultimate origin of a disorder like type II diabetes? At the liver, which pumps out an excessive amount of sugar when it becomes unable to properly respond to insulin? At the muscles? They, too, cannot hear insulin's message and stop picking up glucose from the blood. Or is it perhaps the pancreas, which starts out by producing lots of insulin and then gets tired and stops? How about the abundant adipose tissue that discharges too many fatty acids in those individuals who are on their way to diabetes, or the brain, which controls metabolic pathways through its countless connections? One shouldn't even discount the microbes that inhabit our gut, since they, too, have been implicated in the pathogenesis of diabetes.

Type II diabetes is definitely not the only disorder raising these issues. Interpretation of the pathogenesis of chronic conditions that harass modern societies—including cardiovascular, neurodegenerative, autoimmune, and neoplastic diseases—faces the very same conundrum. An example is Alzheimer's disease, the most famous form of amyloidosis, which started out being considered as a disorder specifically located in the brain but is now recognized as a systemic alteration with major metabolic, cardiovascular, and inflammatory components.

Rather than searching for a single, ultimate cause and attempting to locate a precise physical space, a theory of unbalanced humors may therefore be a more appropriate tool to understand disorders like type II diabetes, Alzheimer's, and several other modern maladies. Perhaps, instead of striving to identify the exact molecule that exerts a precise activity specifically on a single type of cell, a framework that incorporates some degree of ambivalence and ambiguity may help make sense of these kinds of disorders.

The science of body messages such as IL-6 and acute-phase proteins, which straddle artificially partitioned fields of research, seems particularly apt to this goal. Here are mediators that not only challenge the assumption that one molecule can carry only a single message, as multifunctional IL-1, TNF, and other cytokines have done. The message of a mediator like IL-6 can actually change meaning depending on which molecular company that cytokine is keeping. On top of that, the content of the IL-6 message can flip depending on whether this cytokine finds a cell willing to listen (via classical signaling) or one that should be forced to (via trans-signaling). Indeed, it looks like target cells can interpret the message delivered by the same molecule in very different—even opposite—ways. With a cytokine like IL-6, therefore, we may need to change perspective, realizing that the message carried by one molecule does not necessarily have a consistent meaning, that ambiguity is not only possible, but very likely.

In the incomparable words of physicist David Bohm, "I think people get it upside down when they say the unambiguous is the reality and the ambiguous is merely uncertainty about what is really unambiguous. Let's turn it around the other way: the ambiguous is the reality and the unambiguous is merely a special case of it, where we finally manage to pin down some very special aspect."[52]

7

The Fat Revolution: Adipokines

IN THIS LAST chapter, the border between inflammation and metabolism grows even fuzzier as we dive into fat.

About a quarter of the body of a young lean woman (a bit less in men) and more than half of the weight of an obese person—often way more than that—is made of fat, that is, of adipose tissue. As we are somewhat dismally aware, adipose tissue expands with age even in those individuals who manage to remain lean throughout their whole life, resolutely wanting to locate itself exactly in those places where we would rather not see any of it. Yet adipose tissue, which affects the functioning of our organism in so many important ways while also profoundly influencing how we perceive our fellow human beings and ourselves, is rarely at the center of our attention as an essential component of our body. Instead, fat is something many people nowadays would like to get rid of, or move around to locations considered more becoming. But having an appropriate amount of adipose tissue is essential for health, as both too little and too much fat can create multifold issues.

Adipocytes, the main cellular constituents of adipose tissue, produce a bounty of body messages, collectively called adipokines, which play fundamental roles in maintaining homeostasis of the whole organism. After summarizing the social perception and biological function of adipose tissue, in this chapter we will examine the two best-known adipokines, leptin and adiponectin.

The Cultural Meaning of Adipose Tissue

The total amount and the specific location of adipose tissue are major determinants of how we assess the beauty and desirability (and often, lamentably, the worth and status) of human beings, both when we look at other people and when we examine ourselves. Without intending to minimize the emphasis placed of skin color as one of the central elements in the supposed distinction among "human races"—and therefore in the genesis and perpetuation of racism—the manner in which fat is distributed throughout the body adds an additional layer that goes even deeper than skin, both literally and figuratively, in the process of classifying and ranking populations, and individuals within a population.[1]

The location of adipose tissue in the body is in fact a major contributor to anatomical differences that have historically been used to distance and ridicule various peoples and to support the abhorrent notion of their supposed inferiority to Caucasians. Examples abound, and include the steatopygia, that is, accumulation of fat in the buttocks and upper thighs, of women from various parts of Africa (the most [in] famous being the so-called Hottentot Venus) and the derogatorily termed "chinky eyes" of East Asians, caused in large part by the presence of a fat pad in the upper eyelid.[2] Beyond race, fat distribution is one of the major elements in determining whether a particular silhouette is considered masculine or feminine, the size and shape of breasts, bottom, and waist for the most part being determined by the places where adipose tissue prefers to accumulate. Understanding the implications of the cultural norms that determine acceptance or refusal of a given body shape in a society—particularly as they affect women and girls—is a very active field of social inquiry. Aging also has considerable effects on adipose tissue distribution: as men and women get older, fat seemingly migrates from beneath the skin of arms, legs, and face (subcutaneous adipose tissue) to the belly's insides (visceral adipose tissue), changing the body's overall shape and removing support for the skin, leaving it to hang loose and wrinkled.

Besides the characteristics of its location, the total volume of adipose tissue also has salient cultural and societal implications. Long considered a sign of wealth, being overweight or obese has more recently become a stigma in many parts of the world, associated with both accusations (You are nothing but a gluttonous sloth!) and feelings of guilt and shame (I am unable to control myself!). There is little doubt that obesity increases the risk of developing a whole range of pathologies and that strong measures to prevent its further spread should be implemented worldwide. However, an unacceptable contempt toward the obese has come to supplement older forms of discrimination, especially in rich countries where being trim and fit has shifted from being a sign of poverty to a true status symbol. Societies that emphasize thinness are also the most affected by anorexia nervosa, a condition that mostly, though in no way exclusively, affects young women from affluent countries and that stands at the other end of the adiposity spectrum from obesity, and yet is part of the same discourse. People with anorexia nervosa see themselves as too fat even when they are extremely thin. By refraining from ingesting a sufficient amount of calories to sustain the body's metabolic demands and by engaging in behaviors that lead to excessive energy consumption, persons with anorexia nervosa end up at grave risk of death. Wherever success, beauty, and fame are equated with thinness, young women (and some men) can fall into a deadly trap characterized by a disproportionate control over the fundamental survival mechanism of hunger, while many others are shamed for their supposed lack of control over how much they eat.

The Biological Meaning of Adipose Tissue

With this heavy cultural baggage, it is surprising that until quite recently adipose tissue was not considered a worthwhile topic of scientific investigation, or at least not a fashionable one. However, epidemiological recognition of sharply increasing rates of overweight and obese individuals in Westernized countries in the mid-1990s accompanied, and perhaps even spurred, critical discoveries in the field of body messages, contributing to change scientists' perspectives. As a result, in the last quarter century there has been an explosion of research on adipose tissue and the body messages its cells produce.

One of the most important biological functions of adipose tissue is to store energy in the form of triglycerides, which are nothing but a type of fat molecule. Through a series of tightly regulated chemical conversions mediated by enzymatic chains, a portion of the food we eat is transformed into triglycerides that are stored inside adipocytes, the specialized cells of adipose tissue that accumulate fat. Animals, which generally need to move, store energy mostly in the form of triglycerides, a very calorie-dense (nine calories per gram) but lightweight, and therefore easily movable, form of energy. In contrast, plants mainly stash away energy as carbohydrates (mostly starch) that only provide four calories per gram and are relatively heavy compared to fat. Of course, some plants also accumulate energy as fat, particularly in seeds, hence the existence of vegetable oils, while animals have a limited supply of energy saved as carbohydrates in the liver and muscles in the form of glycogen that can be mobilized quickly for immediate use, but is also rapidly depleted.

Triglycerides stored in adipocytes are used between meals and during physical activity to maintain an adequate, long-term energy supply for the whole organism. In addition to acting as an energy buffer between meals, stockpiling of triglycerides inside adipocytes has also been essential during the long history of animal and human evolution, when limited food availability was (and in many places still is) a relatively common phenomenon. It is in fact likely that being able to put away the excess calories that are available during periods of plentiful supplies and use them when resources become scarce has been a crucial aspect of our evolutionary history. However, this particular feature turned into a burden in our modern society of easy access to copious calorie-dense foods and reduced need of physical activity, a situation that presumably contributed to the high rates of overweight and obesity around the world. An additional evolutionary reason for development of a tissue made of specialized fat-storing cells has been that stockpiling of intracellular fat is toxic to most other types of cells—those of the heart and pancreas, for instance. Thus, adipocytes look like a practical and sensible solution to the problem of having to store energy while preventing damage to tissues that are intolerant to fat buildup.[3]

A complex network of signals fine-tunes accumulation of triglycerides inside adipocytes as well as their removal and subsequent use by various cells and tissues. These signals are coordinated with those that regulate the metabolism of carbohydrates and proteins and with the need to provide energy for muscle contraction and countless additional physiological functions. A series of enzymes controlled by the nervous and endocrine systems modulates deposition of fat inside adipocytes (lipogenesis) and its removal (lipolysis). Although some of these enzymes have been discovered only recently, scientists have grasped the regulatory pathways that control the amount of triglycerides stored inside adipocytes for more than half a century. For example, the initial observations that the hormone and neurotransmitter adrenaline induces lipolysis and that insulin instead blocks this pathway of energy mobilization date back at least to the 1950s. However, the discovery in the mid-1990s of two body messages produced by adipocytes—the adipokines leptin and adiponectin—revolutionized our understanding of adipose tissue biology.[4]

As a review of the history of the identification of leptin and adiponectin will illustrate, researchers discovered these two adipokines using diametrically opposed theoretical and practical approaches. Leptin was located using the traditional function-first strategy, after a long chase for the body message that regulates hunger. In contrast, it was a structure-first methodology that drove scientists to adiponectin, with the consequent slow path to understanding its function that we have become familiar with for mediators discovered using this approach. However, the unique and highly selective location of adiponectin's production—adipocytes—provided critical clues to this adipokine's potential function.

Leptin

The year 1994 marks the official birthday of leptin, identified by Dr. Jeffrey Friedman and his collaborators at the Rockefeller University in New York City.[5] However, there is a long history behind this crucial discovery, a history that can be traced over more than a century of biomedical science and that demonstrates the importance of solid systematic research that builds upon its own findings but also illustrates how sagacious, persevering investigators can harness and exploit the vagaries of random luck.

Early Observations

Although leptin is produced by adipocytes, the path leading to its discovery started in the brain, specifically in the hypothalamus, an area containing centers that control several fundamental body functions, including appetite—the instinctive desire to eat.

Clinical and autopsy reports dating from the second half of the 1800s suggested that damage to the hypothalamus, generally caused by tumors or infections, was occasionally associated with rapid, otherwise unexplained, development of obesity. This was surprising, particularly if one considers that a person's body weight is generally maintained inside a relatively narrow range, remaining stable over long periods during early adulthood, then often slowly increasing with aging and eventually declining again in the elderly. This truly remarkable ability to keep weight nearly constant in the face of widely varying access to food and major changes in physical activity has fascinated scientists for a long time, and continues to do so in our ever-more-obese world, where the empty calories we ingest deregulate control mechanisms that were once robust and our contemporary society managed to cripple. In the face of this cunning ability to maintain a relatively stable weight over decades, the sudden massive increase in adiposity observed by nineteenth-century physicians in patients with hypothalamic lesions meant that something was very wrong. However, though decidedly suggestive, those early observations could not demonstrate cause-effect relationships between damage to the hypothalamus and obesity.[6]

During the first half of the twentieth century, several investigators examined this issue, showing that experimentally induced surgical lesions in the hypothalamus of lean rats (and other species)—particularly in an area called ventromedial hypothalamus—led to subsequent development of obesity. As soon as the rodents recovered from the anesthesia necessary to perform brain surgeries, those with hypothalamic lesions started ingesting copious amounts of food and rapidly became obese, whereas control animals that had undergone fake surgeries remained weight-stable. These studies established the scientifically necessary demonstration that lesions in the ventromedial hypothalamus caused the effect of increasing intake of food to such an extent that the animals quickly became massively obese, overcoming what-ever message saying "stop eating" or "do not start eating again" an intact hypothalamus was generally able to sense, a message that can thus be defined as carrying a signal of satiety, of having ingested a suf-ficient amount of calories.[7]

The next obvious step was to ask how this might happen. Nobody had any clue as to the nature of the supposed message conveying the satiety signal to the hypothalamus. Two major theories emerged. One placed digestion of food at its core, postulating that the heat released when food was broken down into its elementary components directly affected the hypothalamus (this brain area regulates both body tem-perature and appetite, so the idea of heat made sense). The second theory, advanced in 1952 by British researcher Gordon Kennedy, pro-posed that some sort of mediator coming from adipose tissue informed the hypothalamus of exactly how much fat was present in the body, conveying to the brain a message saying "stop eating, there are enough energy reserves to keep you going, you can safely direct your attention to tasks other than ingestion of more calories." This came to be known as the lipostat, or lipostatic, theory, the postulation of a "stop eating" message, a satiety message, derived from adipose tissue.[8]

Shortly thereafter, another British researcher, the physician-scientist Romaine Hervey, set out to demonstrate the validity of the lipostat theory using an experimental technique, called parabiosis, that was instrumental in moving the field forward both at this stage and later on. Parabiosis (the term means living beside) is the anatomical union of two organisms, something that occurs naturally in conjoined twins but a condition that an investigator can also induce surgically in exper-imental animals. I must honestly admit that discussing parabiosis makes even a seasoned scientist like myself, who has spent decades performing research with laboratory rodents, rather uneasy. If you are picturing the image of an otherworldly furry creature with two heads, eight feet and two tails you are not that far away from the reality of this type of experimental approach. In parabiotic experiments, in fact, the skin and some of the muscles on the flanks of two rats or mice (the most commonly used species for these studies) are stitched together, with the ultimate goal of making the two animals share their blood into what eventually becomes a common circulatory system.[9]

Two animals joined in parabiosis mix their blood, thus allowing a body message produced and released by the cells of one animal in the pair to reach and activate its specific receptors present on the cells of the surgically connected partner. What kind of pivotal experiments did Dr. Romaine Hervey perform using parabiosis? He surgically connected pairs of rats, made sure there was indeed the expected exchange of blood between the two rodents, and then lesioned the hypothalamus of one of the animals in the pair, leaving the second rat with an intact brain. Control groups consisted of pairs of parabiotic rats without any hypothalamic lesions as well as free-living single rats with or without brain lesions. Dr. Hervey's 1959 article confirmed previous data demonstrating that rats with a damaged hypothalamus rapidly became obese due to excess feeding, reaching a point where about half of their weight was made of adipose tissue, just like obese humans. However, the true novelty of Hervey's experiments consisted in demonstrating that the intact parabiotic partners of brain-damaged rats—the partners without hypothalamic lesions—rather than becoming obese instead completely lost interest in eating and became very thin and emaciated. These rats would not eat even if food was directly placed in front of them and some even ended up dying of starvation.[10]

These results were interpreted by Hervey and his colleagues as indicating that the adipose tissue of the rat that had become obese due to damage to its hypothalamus released high levels of a satiety factor that entered the common bloodstream of the parabiotic pair and was sensed by the intact hypothalamus of its partner, which stopped eating and thus starved to death because its brain was receiving bogus information about the state of its own energy reserves. In contrast, the lesioned hypothalamus of the rat that had undergone brain surgery could not sense this putative satiety message. Therefore, this rat kept eating, becoming heavier and heavier while releasing ever more of the satiety message its own brain could not perceive. By being unable to sense this message, the brain of the rat with hypothalamic lesions came to the conclusion that the animal was starving and thus pushed the hunger accelerator pedal: the rat could not stop eating. In contrast, the intact hypothalamus of the non-lesioned rat was flooded with the satiety message coming from the expanded adipose tissue of its connected partner, information its brain incorrectly interpreted as having accumulated too much energy in the form of fat. This time it was the hunger break that was pushed rather than the hunger accelerator, and this rat would not eat no matter what. Paradoxically, thus, the hypothalamic-lesioned obese partner thought of itself as being too thin and thus kept on eating, while the hypothalamic-intact thin partner considered itself massively obese and would therefore not eat.

These results, and additional data later obtained by Dr. Hervey and other investigators, seemed to strongly support Gordon Kennedy's lipostat theory: as adipose tissue expanded, it produced a message telling the hypothalamus to go easy with calorie intake. However, the identity of the "do not eat" message remained elusive.

Fat Mice

Let us now change countries, from the United Kingdom to the United States, and move back in time. We are at the Jackson Laboratory, a biomedical research center located in Bar Harbor, on the wooded coast of Maine. Founded in 1929, the institute was the brainchild of Clarence Cook Little, a geneticist who had then just ended his term as president of the University of Maine. During his scientific career, Little had made significant contributions toward adopting mice as experimental animals for biomedical research, particularly in the field on cancer. Now he wanted to build a place dedicated to raising and examining laboratory mice but also distributing them to the scientific community to support other investigators. Dr. Little managed to obtain funding from a group of Detroit industrialists, including Edsel Ford, who presided over the Ford Motor Company, and Roscoe B. Jackson, the president of the Hudson Motorcar Company, after whom the institute was named. In 1933, the center began shipping mice to investigators located at other institutions. It has since grown to be not only one of the largest providers of laboratory mice, but also an outstanding research center, its staff expanding from the initial seven members to more than a thousand.[11]

In 1947, a major fire destroyed many of the buildings at the Jackson Laboratory and killed off most of the mouse colonies housed there. Donations of both funds and mice poured in from all over the world to help repopulate the lost colonies, a gesture that becomes even more significant if one considers the historical period in which it took place, just two years after the end of World War II. These donations permitted operations to restart in record time, so that a rebuilt and restocked institute could reopen in 1948.[12]

Only one short year after the institute's reconstruction, a one-page article published in the *Journal of Heredity* laconically announced, "In the summer of 1949 some very plump young mice were found." A new mutation, called *obese* and designated as *ob*, had appeared in a mouse colony at the Jackson Laboratory, with mutant mice rapidly becoming obese, reaching a body weight about four times larger than that of their normal counterparts. Randomness had acted on the DNA of a mouse bred at the Jackson Laboratory, generating a genetic change that made that mouse and its progeny eat at a frantic pace, thus becoming extremely fat. Whether this mutation happened in one of the rodents that had survived the 1947 fire or one donated by a research facility, perhaps on the other side of the world, we will never know.[13]

During subsequent years, various investigators studied *ob* mice and obtained extremely valuable information. Careful examination demonstrated these animals suffered from the metabolic abnormalities you would expect given their extreme obesity, such as altered glucose and fat metabolism and a massively enlarged, fatty liver. A series of puzzling observations also began to emerge. Mice with the *ob* mutation were infertile and had reduced activity of the sympathetic nervous system. Their bones were more fragile than they should have been and parts of their immune system were defective, while their liver looked much worse than expected.[14] However, despite multiple meticulous studies on various aspects of the physiology of *ob* mice, nobody could get at the root of what exactly was wrong with them.

As if one stroke of random luck were not enough, a second obese mouse appeared at the Jackson Laboratory about fifteen years after the *ob* mutation had established itself. In 1966, Dr. Douglas Coleman, a biochemist who had joined the Jackson Laboratory a few years earlier, revealed that a new type of obese mouse had surfaced. Coleman and colleagues called this new variant *diabetes*, or *db*, since these new mice were not only massively obese, like their *ob* cousins, but also severely diabetic.[15] Coleman spent most of his subsequent working life characterizing mice with *ob* and *db* mutations, using parabiosis as a powerful tool to explore the mechanisms underlying these animals' obesity and diabetes.

In a series of experiments, Dr. Coleman and other investigators connected an *ob* mouse to a normal mouse through parabiosis, demonstrating that the *ob* mouse ate less and became less obese than it would have had the researchers left it alone, whereas not much happened to its attached, nonmutant normal partner. This result suggested that a blood-borne satiety message coming from the normal mouse somehow informed the *ob* mouse to reduce its intake of food.

In a separate study, Dr. Coleman generated pairs of *db* mice connected to normal mice. The result of this type of pairing matched the data first obtained by Romaine Hervey through parabiosis of rats with hypothalamic lesions: normal mice in these pairs lost interest in food and died by starvation, whereas no effects were apparent in their *db* partners. Similar to the rat studies of the late 1950s, Coleman and his colleagues interpreted these experiments according to the lipostat theory: just as rats with hypothalamic lesions were incapable of sensing the high levels of satiety factor generated by their own adipose tissue, the result of the new parabiosis experiments suggested that the *db* mutation made mice unable to respond to their own putative adipocyte-derived satiety factor. However, the hypothalamus of their parabiosed partners could properly sense the elevated amounts of satiety factor that were pouring in from the surgically connected *db* mouse. As a result, these mice starved themselves to death as had happened to Romaine Hervey's rats.[16]

These were all powerful results that strongly supported the lipostat theory, but the real breakthrough came in 1973 when, in a famous article, Dr. Coleman suggested the existence of a very intimate connection between the *ob* and *db* mutations. The key experiment in this report was the generation of parabiotic pairs consisting of one mouse with the *ob* mutation and one *db* mutant, both of whom would become massively obese if left alone. The most notable finding was that the *ob* mouse in the pair stopped eating, dramatically lost weight, and died by starvation, whereas its *db* partner kept on chewing its food and getting bigger and bigger. "Since *ob* mice in parabiosis with *db* mice respond like normal mice it can be concluded that the *ob* mice have normal satiety centers sensitive to the satiety factor produced by the *db* partner," Dr. Coleman concluded. Showing remarkable ingenuity in interpreting the meaning of his experimental findings, Coleman proposed that the satiety center of *db* mice would be defective, unable to respond to the satiety message, whereas *ob* mice would be incapable of producing the very same satiety factor but perfectly able to respond to this molecule when placed in parabiosis with *db* mice, whose cells generated high amounts of satiety message in an attempt to overcome their hypothalamic inability to sense it. "Both mechanisms would cause over-eating and produce identical changes in both *ob* and *db* mice," Dr. Coleman deduced. His prescient explanation precisely predicted what would take another two full decades to be formally demonstrated.[17]

Coleman's seminal article was published in 1973, when—as we discussed in previous chapters—analysis and manipulation of genetic information was still in its infancy. It was also a time when the obesity epidemic had not yet materialized, and therefore an investigation trying to understand how and why individuals might put on extra body fat was not necessarily top priority. Consequently, even though scientists knew that *ob* and *db* mice were obese due to alterations in their DNA, both technical and programmatic issues prevented identification of the affected genes.

Locating the Satiety Message

We need to change time and place yet again. We are now in New York City in 1986 when, after obtaining his medical degree, Jeffrey Friedman also earned a PhD and an appointment as assistant professor at the Rockefeller University and the Howard Hughes Medical Institute, a science philanthropy that provides substantial funding to exceptionally promising and accomplished investigators. Fascinated by the genetics of complex behaviors, particularly eating, Dr. Friedman decided he would be the one to finally locate the cause of obesity in *ob* mice by identifying the gene whose mutation Jackson Laboratory researchers had serendipitously noted almost four decades earlier. He would identify the satiety message whose existence Gordon Kennedy had postulated in the 1950s and Douglas Coleman had painstakingly investigated using *ob* and *db* mice in the 1970s.

From the moment they started to work on this specific question, it took Friedman and his collaborators more than five years and almost one thousand mice to track down the exact position of the *ob* gene through a technique called positional cloning. Using the inheritance of known genes as landmarks, the researchers diligently narrowed down the stretch of DNA in which they thought the *ob* gene would be located, until they ended up with a section containing only six genes. At this point, they checked where these six genes were turned on, separately analyzing tissues and organs to evaluate the precise location in which the DNA of these genes was transcribed into messenger RNA and therefore translated into proteins. If you recall that the lipostat theory postulated adipose tissue would be the source of the satiety message, it will come as no surprise that Friedman excitedly described as a eureka moment the time when he held in his hands visible proof that one of those six genes was exclusively turned on precisely in adipose tissue. "Overnight I had set up an experiment and I developed it the next morning at five or six, because I could not sleep. The result was very striking: when I saw it, I instantly knew that not only we had cloned the *ob* gene, but that Coleman's hypothesis was likely to be correct."[18]

In December 1994, Friedman's team published the sequence of the *ob* gene, the genetic code for the long sought-after satiety message postulated by the lipostat theory. In the same article the investigators also identified the mutation that inactivated the message in *ob* mice as the change of a single letter in the DNA sequence (from a C to a T) that compellingly transformed the meaning of the code from an instruction to insert the amino acid arginine into the command to stop transcription of the gene, therefore leading to production of a truncated RNA that could not be turned into a functional protein—what is known as a nonsense mutation. Critically, Friedman's group also reported the sequence of the human equivalent of the mouse *ob* gene, thus opening the road to understanding the control of appetite and body weight in humans.[19]

Research in the field veritably erupted, with the importance of these findings also felt strongly outside the world of academia. Under different circumstances, interest for even such a remarkable breakthrough would have been restricted to the scientific community. By the mid-1990s, however, obesity rates had started to skyrocket in the

United States and other developed countries, turning identification of the fat hormone into a true darling of the lay media and the biotech industry. Whereas putting adequate food on people's tables was a main concern in the 1950s, when the lipostat theory had been originally formulated, and malnutrition was still a dominant problem in the early 1970s when Douglas Coleman published the results of his parabiosis experiments, things had turned around in the developed world by 1994–95. Now, confronting the deleterious effects of calorie over-consumption had become a major issue, even in the face of lingering, still far-reaching dietary deficiencies. Thus, the time was ripe for society to hear about the amazing gene that controlled how much we eat, and for industry to grasp the potential for developing this finding into practical, potentially lucrative applications.

In July 1995, only a few months after Dr. Friedman's original article, three reports appeared in the same issue of the journal *Science* demonstrating that injections of recombinant OB protein reduced appetite and induced weight loss in mice. Although one of the three papers was from Jeff Friedman's group, the other two came from studies done outside of academia, at Amgen, the California biotech, and Hoffmann-La Roche, a more traditional pharmaceutical company. In their paper, Friedman and colleagues utilized the Greek root *leptos*—which means thin—to coin the new term leptin for the satiety message they had cloned, a term that would become so famous it yielded more than six million hits on a Google search two decades later.[20]

Identification of the receptor for leptin, the gene hypothesized to be mutated in *db* mice, soon followed, with an article in December 1995 by Millennium Pharmaceuticals (a biotech company of which Friedman was a founding member), tailed in 1996 by additional reports. These investigations and subsequent ones, to which Dr. Rudolph L. Leibel substantially contributed, not only identified the gene coding for the leptin receptor but also demonstrated the cause of obesity in *db* mice as another single-letter DNA change, one that rendered the receptor for leptin nonfunctional, just as Douglas Coleman had postulated. These studies and other work that rapidly followed also indicated that the leptin receptor was present at high levels in the hypothalamus, confirming the lipostat hypothesis proposed by Dr. Kennedy and elaborated by Dr. Hervey in the 1950s.[21]

Many researchers have since investigated the neural pathways that are modulated when leptin reaches the hypothalamus and other brain areas. Specifically, leptin activates a family of neurons that are called anorexigenic because they reduce the sensation of hunger, while it concurrently suppresses the activity of neurons that are instead called orexigenic because they increase hunger. These terms, orexigenic and anorexigenic, derive from the Greek root *orexis*, which means desire and, by extension, appetite. Leptin activates anorexigenic neural pathways, those that make you decide not to attack that last piece of pizza because you feel you have already had enough food, you have reached satiety. While stimulating neurons that make you stop eating, leptin concomitantly also decreases the activity of the orexigenic pathways, those that would tell you to go ahead and gulp down that extra slice.

But that's not all. Investigators have shown that leptin regulates additional neural pathways. One of the scientists who performed pivotal work in characterizing the neurobiology of leptin is Dr. Martin Myers, Jr., professor of diabetes research at the University of Michigan in Ann Arbor. Myers was born when his own father was in medical school, the family initially moving around the country while Dr. Myers Sr. completed his training and then settling on a farm just outside Iowa City. They then picked up again and moved to Cincinnati when Martin Jr. started high school. While in college at Princeton University, Myers didn't know whether he wanted to study biomedical or political science. "The thing that really turned me on was a molecular biology course that I took as a freshman," Myers explained during our interview. "Unlike the biology I had seen in high school, this wasn't just memorization but rather taught the essentials of how we know about DNA by reviewing the classic experiments in the field. I wasn't sure whether I wanted to do research or be a physician, so I enrolled in the MD/PhD program at Harvard University. I got into a lab early on, working with Morris White whom I had met as a medical school tutor. I loved research on insulin signaling in Morris's lab, whereas clinical medicine, while enjoyable, wasn't nearly as exciting. Thus, after medical school I went right back to the lab instead of doing clinical training. When I graduated from the MD/PhD program, I was fortunate to have the opportunity to set up my own lab at the Joslin Diabetes Center in Boston," Dr. Myers continued. "Ron Kahn, the head of the Joslin, told me that I would have to move away from insulin signaling and find something new to make my own. I was thus casting about looking for an interesting problem that involved signaling and was relevant to metabolism. This was just after the leptin receptor had been cloned, so I jumped on that."[22]

Since that receptor had just been cloned, little was known about it and there was a lot of basic, fundamental research to do. Myers and colleagues spent several years working out the details of leptin receptor signaling in cells *in vitro* and then moved to mouse models *in vivo*. Using genetic engineering, they disrupted specific pathways activated by leptin and analyzed the fine-tuned involvement of each pathway in regulation of appetite and energy expenditure. They then went on, exploring the role of leptin signaling outside the hypothalamus. "These other non-hypothalamic neurons comprise about 80–90 percent of all leptin receptor-expressing neurons, so they likely contribute importantly to leptin's action," Dr. Myers explained. "Several sets of neurons that we are particularly excited about include those that control the reward system and those that contribute to the control of glucose homeostasis." In fact, the vast majority of neurons that respond to leptin are part of what is known as the reward system, which determines the sensations of pleasure we get from eating. By keeping this system in check, leptin reduces the incentive to eat. Consequently, on top of not feeling that hungry and of getting easily full (responses that involve modulation of hypothalamic pathways), when leptin concentrations are elevated and brain leptin receptors are activated we are also less attracted to food as an item of pleasure. Things are of course more complicated than this, with several additional factors contributing to the decision to begin, continue, or stop eating. Despite being just one among these many factors, leptin is fundamental, as demonstrated by the extreme phenotype of mice that cannot produce or sense this adipokine—*ob* and *db* mice—and that were instrumental in its discovery.[23]

In 2010, Drs. Coleman and Friedman jointly received the Albert Lasker Basic Medical Research Award for the discovery of leptin, a prestigious and well-deserved recognition. The uncommon foresight of both the Rockefeller University and the Howard Hughes Medical Institute, which kept funding Jeffrey Friedman throughout his long-term project even though success was definitely not a given, should also be praised for the discovery. In hindsight, this turned out to be a good investment both in terms of scientific recognition and, as we shall see next, financial return.

A New Treatment for Obesity?

With obesity rates on the rise and a gene finally at hand, the biotech and pharmaceutical industry jumped at the opportunity to exploit leptin for development of antiobesity drugs. Immediately after identification of the *ob* gene, fifteen companies expressed interest in obtaining the rights to develop products based on this discovery, with Amgen eventually winning the bid. A news article published in *Science* in May 1995, before the name leptin had even been publicly reported, discussed how Amgen was going to pay $20 million for the exclusive license to develop drugs based on the OB protein. Reportedly, this represented the largest up-front payment ever for rights to a university patent.[24]

That huge financial deal quickly spurred controversy due to improper handling by an analyst working for the then thriving, and since disgraced and defunct, financial company Lehman Brothers. The analyst had disclosed news to several thousand clients about the impending publication of articles demonstrating that leptin could slim mice, thus triggering a surge in Amgen's shares. At the time of this improper revelation, the articles showing leptin's ability to induce weight loss in mice were still under embargo; accusations of differential disclosure, of providing some investors with information not available to others, ran rampant. However, as we are about to appreciate, these concerns may have been irrelevant, since the prospect of turning that initial $20 million investment into lofty profits soon started to look substantially dimmer for Amgen and its stockholders.[25]

With unprecedented speed, in 1996 Amgen began a trial to test the safety and tolerability of recombinant leptin in humans, and in 1997 launched additional studies to evaluate further this mediator's effectiveness for treatment of obesity and type II diabetes. Although some obese subjects enrolled in these trials lost a considerable amount of weight when injected with the highest doses of leptin, results were quite variable and overall not satisfying enough to justify the additional major investment needed to run the large clinical trial that would be necessary to definitively prove leptin's worth in treating obesity or diabetes. It didn't take much else for Amgen's leadership to decide to quietly ease their pricey acquisition into a hushed sleep and direct their attention to other issues.[26]

Rare Causes of Obesity

Why were Amgen's clinical trials coming up with results so lukewarm as to discourage further investments in leptin despite the big money spent to acquire the initial rights to its development? After all, leptin had shown significant promise just a few years earlier. To understand the reasons for such a desultory ending, we need to leave behind the financial world and return to biology.

As just reviewed, the phenotype of *ob* and *db* mice was the basis for identification of leptin and its receptor. Mice with a mutated *ob* gene are obese because they cannot produce the satiety message that would tell the hypothalamus "stop eating—you have enough fat stored in your adipose tissue." On the other hand, *db* mice are obese due to an altered leptin receptor gene that prevents their hypothalamus from responding to leptin, which is produced in such abundance that adipocytes of *db* mice keep yelling "stop eating" to a hypothalamus that is unable to hear even the loudest of messages. Similar genetic alterations have been discovered in strains of rats. More poignantly, mutations in either the leptin or leptin receptor gene are a cause, even if extremely rare, of obesity in people.

The first changes that inactivate the leptin gene in humans, the equivalent of the mouse *ob* mutation, were identified in 1997 and 1998 in members of families of Pakistani and Turkish origin who were not only massively obese but also for the most part infertile, some of them never having reached puberty despite being in their twenties and thirties. In exact parallel to what had happened to *ob* mice and had sparked the commercial interest in leptin, injections of this body message dramatically changed the life of individuals with mutations in the leptin gene, replacing the missing satiety message and thus taming their overwhelming hunger drive, as well as providing the required stimulus for appropriate production of sex hormones and correction of infertility. Thus, leptin has truly been a lifesaving intervention for people who are obese due to lack of leptin production. Unfortunately, this is not the case for persons with mutations that inactivate the leptin receptor gene, the human counterpart of the mouse *db* mutation.[27]

One of the scientists who contributed most to both scientific research and clinical care of individuals carrying mutations in the leptin receptor gene is Dr. Karine Clément, currently director of the Institute for Cardiometabolism and Nutrition in Paris, France. Clément was born in Lille, in the north of France, her father a professor of geology, her mother a teacher of English. As an adolescent, Karine never thought she would become an academic scientist, wanting instead to be a physician, perhaps a pediatrician or a psychiatrist. She went to medical school and moved to Paris in 1989 for her medical internship. After a start in pediatric diseases, she spent six months in the Department of Nutrition at Hôtel-Dieu de Paris. This hospital department was quite unique in France, being fully dedicated to the care of human obesity, particularly the most extreme cases. In the 1970s–1980s, the two heads of the department—Bernard Guy-Grand and Arnaud Basdevant—had pioneered the concept that obesity care needed a multidisciplinary clinical approach, with physicians, nutritionists, psychologists, experts in physical activity, and (later on) surgeons. "I became immediately interested in this field because of its multifaceted dimensions, including psychological, social, environmental, and many unknown biological and complex aspects," Dr. Clément recounted during our interview. In order to obtain research experience, during the internship she enrolled in a master's of science program and worked at the Center for the Study of Human Polymorphism, specifically focusing on a form of diabetes known as MODY (maturity-onset diabetes of the young), a relatively rare form of diabetes caused by mutations in a single gene. "This was a terrific time to be at the center," Dr. Clément continued, "as it coincided with the discovery of gene mutations in families with members affected by MODY by the research team headed by Professor Daniel Cohen and Dr. Philippe Froguel. Therefore, I had the opportunity to contribute to clinical investigations exploring glucose homeostasis in patients bearing these mutations. I learned a lot about the genetics of metabolic

diseases and metabolic research in general. As a result of this experience, I decided to do a PhD in parallel with medical school. I guess I
have been very lucky to be part of this work and contribute to these
discoveries."[28]

Together with colleagues, Dr. Clément thought she could perform
genetic studies similar to those that had taken place on patients with
MODY but studying obesity instead of diabetes. She therefore focused
her PhD dissertation on the genetics of extreme obesity. She completed
her medical education while working as a fellow in the Department of
Nutrition at Hôtel-Dieu, where the team she was part of discovered
mutations in the leptin receptor gene that caused massive obesity. As
Dr. Clément explained, "We had in the clinical department a seven-
teen-year-old patient with extremely severe obesity (more than 150
kg), pronounced food-seeking behavior, and no puberty. I also started
to follow her thirteen-year-old sister with the same phenotype (140 kg
at that time). They were from a consanguineous family with nine children, and three sisters had the same obesity issue. Because measuring
serum leptin became available around that time, we measured it in the
sisters' blood with a biologist colleague [Dr. Najiba Lahlou] and discovered that levels of leptin were very high, more than tenfold higher
compared to subjects with similar fat mass. Thus, with my close colleague Dr. Christian Vaisse, we started to study the leptin receptor
gene and discovered a mutation."[29]

What Dr. Clément and her colleagues had discovered was that these
sisters had a change in their DNA that rendered their leptin receptors
nonfunctional; therefore, their hypothalamus was unable to respond
even to the extremely elevated amounts of leptin produced by their
adipose tissue, just as Douglas Coleman had shown in *db* mice many
years earlier. In 1998, Dr. Clément and colleagues published their findings in *Nature*. Since then, investigators have discovered other leptin
receptor mutations, all with extreme obesity of early onset, food impul-
sivity, and pubertal retardation, and all extremely rare.[30]

Over the years, Dr. Clément continued to investigate the causes
and consequences of obesity, though her focus shifted from the rare
forms due to genetic alterations to the more common environmentally
induced type. However, she continued caring for the sisters with leptin
receptor mutations. "For me this discovery has also been a human

adventure with the patients and their family," she told me. "As a doctor, I am still following the youngest patient of the family with the leptin receptor mutation. When she reached the age of eighteen, she asked me if she could have children. This was a very difficult question, because as a physician and scientist starting to learn about leptin biology, I knew that leptin was deeply involved in the reproductive system and that animals with leptin or leptin receptor mutations were infertile. I was cautious enough to tell the patient that I did not know what would happen and that in her medical condition there were a lot of unknown things. The patient got pregnant at twenty-six and delivered a healthy boy, followed by another one a few years later. I was really happy for her. My regret is that we don't have a targeted treatment for this patient and her sisters."[31]

Why is it that physicians do not have treatments for patients carrying mutations of the leptin receptor gene while they can instead successfully treat obesity in persons with mutations of the leptin gene, as mentioned above? The reason is that people with an altered leptin gene lack the satiety signal but are perfectly able to respond to it. Hence, when they receive injections of recombinant leptin the molecule travels through the bloodstream to their hypothalamus and delivers its message, reducing their eating drive and eventually allowing them to lose weight. In contrast, individuals with mutations in the leptin receptor gene are unable to respond to the satiety message even though their hypothalamus is flooded with leptin: giving them more leptin through injections is not going to make them any more responsive. Thus, unfortunately, there are currently no specific treatments for these patients.

This long detour into exceedingly rare genetic causes of obesity has brought us back to the reason for the disappointing outcome of clinical trials that attempted to use leptin to treat common forms of obesity. Although lack of leptin due to mutations that inactivate the *ob* gene results in obesity because the signal that tells the brain to stop eating is missing, the vast majority of obese people actually have lots of leptin in their bloodstream, an observation made shortly after the identification of leptin itself. In fact, in individuals affected by the most common forms of obesity, leptin is present in the circulation at levels that correspond to how much adipose tissue a person has. The equation is rather simple, the more adipose tissue, the more leptin. Therefore, further increasing the satiety message by administering recombinant leptin to someone who is "commonly" obese does not make a lot of sense, since there is only so much signal the brain can perceive and respond to. It seems like simple logic, and yet the pull of a vast and potentially lucrative market in which effective and safe pharmacological treatments for obesity were—and still are—lacking had been strong enough to convince Amgen's leadership to launch expensive clinical trials that, based on scientific knowledge already available at the time, were destined to fail.[32]

Evolution of a Message

The logic behind this reasoning, however, is only deceptively simple. If leptin really is so crucial for regulating appetite, for letting the brain know we have enough energy stored inside our adipocytes, how come leptin's progressive increase as we get fatter and fatter does not stop us from eating excessively and putting on even more weight? In other words, if leptin does indeed carry an important "stop eating" message to the brain and if there is a lot of leptin in the blood of those who are obese, why is it that they keep ingesting more and more calories? Why is it that *ob* mice flooded with leptin through their parabiosed *db* partners died of starvation whereas individuals whose brain is flooded by the leptin that comes from their own expanded adipose tissue become, and remain, obese? Does all the accumulated evidence from parabiosis approaches and from data showing that lack of leptin or a response to it makes rodents obese only represent a meaningless sequence of experiments? If having high levels of leptin does not actually make a difference in the amount of food people eat, how should we interpret the massive obesity of patients with *ob* or *db* gene mutations?

The answer to these questions may be found in our remote history. During evolution, not having enough calories was probably a much greater threat to survival than having too many. As a result, we can expect the defense against starvation—mirrored in the body by either low or rapidly falling levels of leptin—to produce a much stronger response than the defense against obesity, represented by high leptin. In other words, not having enough leptin (which means being too thin and therefore lacking sufficient energy reserves) or having leptin levels that quickly fall (as when someone loses weight, particularly if this happens rapidly) sends an extremely powerful message to the brain, a message that says "eat" with such compelling force as to leave little choice. Having low or falling leptin can thus be conceived as a primordial command that kicks in when energy stores inside adipocytes are either inadequate or rapidly reduced, a message that impels a person to procure and consume food, while shifting metabolic pathways toward energy conservation. Inadequate levels of leptin often also result in infertility, particularly in women, thus preventing the possibility of a pregnancy in an individual that would be unlikely to successfully carry it to term, given the paucity of her fat stores in the face of the onerous energy demands of childbearing.

While these responses have likely been evolutionarily precious to the survival of our species (and all other species that possess similar mechanisms), the opposite cannot be said for the high leptin levels of common obesity. Since being too big has rarely—if ever—been an issue in our past, we presumably did not evolve mechanisms to protect us from having too much leptin. Hence, once our adipocytes produce enough leptin to reach the threshold that signals for the presence of sufficient energy stores, going even higher does likely not make much difference to the brain. This implies that having very little fat is a dangerous situation that requires immediate intervention to ensure survival but, as soon as a bit of energy is stored away in adipocytes to permit synthesis of some leptin, the system calms down, at which point it does not matter anymore how strong the satiety signal might become. Consequently, while in all likelihood evolution gave us a tremendously powerful mechanism that says "not enough," we are left without a way to say "too much." As mentioned at the end of the previous chapter when talking about IL-6 and hepcidin, obesity seems to throw the organism off the exquisite balance achieved throughout its own evolutionary history, clashing with the abrupt and unforeseen dilemma of having to manage an excess of energy reserves.[33]

Having finally listened to the evolutionary lesson, the pharmaceutical industry has more recently decided it would make more sense to develop leptin commercially as a drug that would target conditions characterized by the strong "not enough" message rather than the weak "too much" one. In 2014, the Food and Drug Administration approved a modified version of recombinant leptin, called metreleptin, for treatment of lipodystrophy, a very rare and dangerous condition in which individuals do not produce leptin because their bodies are unable to generate adipocytes. Leptin is luckily an extremely effective and life-changing treatment for these patients, as it is for those harboring mutations in the leptin gene we discussed above, but it is certainly not a blockbuster drug for the industry. By changing disease target from obesity to lipodystrophy and leptin gene mutations, the market in fact shrank from millions of potential customers to, at best, hundreds.[34]

But not all is lost in terms of exploiting leptin's biology to help commonly obese people lose weight. Scientists are testing whether administering leptin after an initial period of weight loss can support further pound shedding and maintenance of the reduced weight, generally the most challenging part of any weight loss program. In fact, leptin levels sharply decline even in the presence of relatively minor body weight loss. Here, it isn't the absolute amount of leptin that matters, but rather the differential from pre weight loss to active weight loss, a response that sends a message informing the brain that something is wrong, that there is an impelling need to correct such a state of negative energy balance. This strong response is possibly a key factor in the failure of many obese people to maintain weight loss: as soon as fat stores start going down, leptin levels abate and the brain screams "eat" with such force that conscious self-control becomes extremely onerous, and in many cases impossible. It's millions of years of evolutionary history distilled into one single body message, no wonder it's so hard to fight it. Better would be never to get to the point of having to lose weight, but here collusion between a biology that pushes us to store energy in our adipocytes and a society that assaults us with access to excess calories and the convenience of a sedentary lifestyle becomes even harder to confront.[35]

Adiponectin

Though its name may rhyme with leptin, adiponectin's history could not be more different from that of its sister adipokine. Whereas identification of leptin stood on top of a century-old accretion of findings pointing to the existence of a satiety factor made by adipose tissue, adiponectin was instead discovered seemingly out of the blue by researchers who had gone hunting for something they had no clue even existed, let alone any idea what it might do. Like those members of the IL-1 family of cytokines identified through *in silico* approaches (reviewed in Chapter 3), and in stark contrast with leptin, the discovery of adiponectin was not the result of attempting to identify the molecular basis of a specific biological function, but rather proceeded in the opposite direction: structure first, function later.

When Time Is Right

Four separate investigative teams, two in Massachusetts and two in Japan, ended up discovering the same adipokine at about the same time, the mid-1990s. This was shortly after the long search for leptin had finally reached its culmination thanks to the accomplishments of Dr. Friedman's group, indicating how ready that era was for new questions about adipose tissue.

As we have seen in previous chapters, it is rather common for separate researchers to discover a body message almost simultaneously through parallel lines of investigation. However, none of the scientists that independently discovered adiponectin was trying to find the substrate for a particular function. Instead, these researchers were all engaged in what can best be described as molecular rummaging. In fact, three of the four groups found adiponectin by combing through genes that would be turned on in adipocytes but not in other cells, while the fourth scrutinized proteins present in blood. A fishing expedition without a solid scientific hypothesis some would disparagingly describe it, yet one that yielded crucial findings in the form of a new adipokine. What the researchers discovered was a body message that— like leptin—was produced exclusively by adipocytes, but one that was present in the bloodstream at incredibly high levels, at least a thousandfold higher than leptin. How could a protein that constitutes a sizable portion of an adipocyte's output and that circulates in blood at astounding concentrations have gone missing until 1995?

"This was the same time when Jeff Friedman's leptin description came out," Dr. Philipp Scherer, first author of the first paper describing the discovery of adiponectin, commented during our interview, hence the intense curiosity about other mediators adipocytes might possibly produce. Born and raised in a small town in the mountains of central Switzerland, Philipp is one of four siblings who grew up in the train station where their father was stationmaster for the Swiss Federal Railways. The family frequently used the train to travel around Switzerland and the rest of Europe. Then, for one year during high school, Scherer was an exchange student near Buffalo, New York. "This was a life-altering experience for me, the time when I discovered and embraced my love for biological sciences," he explained. "I returned to Switzerland for undergraduate and graduate school, both at the University of Basel, where I obtained my PhD in Dr. [Gottfried] Schatz's lab studying proteins' behavior in cell membranes. I had loved my high school chemistry and biology classes, and initially aimed for medical school, but at the last minute decided I preferred research and instead switched to a PhD. I then continued my studies in Harvey Lodish's lab as a postdoc at the Massachusetts Institute of Technology." It is there that Philipp Scherer began focusing on adipocytes and, when the time was right, eventually landed on adiponectin.[36]

The discovery of leptin had stimulated interest in exploring whether adipocytes might produce additional mediators that affected metabolism, a question that wasn't on many scientists' radar before 1994, the year Friedman identified the *ob* gene. When there is no question, there will be no search for an answer. That adiponectin has an extremely complex biology and exerts functions that—though pivotal—can best be described as modulatory of other systems is an additional reason that likely contributed to its late recognition compared to other body messages. In fact, even assuming that chance would decide to play its tricks again by generating the adiponectin equivalent of an *ob* or *db* mouse, that is, should a spontaneous mutation in the gene for either adiponectin or its receptors have surfaced in a rodent colony somewhere, it is unlikely this would have pushed the field forward. Indeed, these putative adiponectin mutants would not have been as easily recognized as *ob* or *db* mice had been, because, even though adiponectin plays fundamental roles in the regulation of metabolism, its absence does not cause disruptions nearly as visible as those originated by lack of leptin or its receptor.

That the time was ripe for the discovery of another body message from adipose tissue is clearly indicated by the tight succession of publications reporting identification of the new adipokine. The first article, by Dr. Scherer and his colleagues at the Massachusetts Institute of Technology, was submitted for publication to the *Journal of Biological Chemistry* at the end of July 1995 and published in November. The second, by Drs. Erding Hu, Peng Liang, and Bruce Spiegelman at nearby Harvard University, was submitted to the same journal less than a month after the first paper, but it took a bit longer to go through the revision process and therefore the results were not made public until May 1996. The third report, by Dr. Kazuhisa Maeda and his colleagues at Osaka University, was submitted in March 1996 and published in April of the same year, while the work of Dr. Yasuko Nakano and colleagues at Showa University was submitted in June 1996 and published in October.[37]

Each of the first three articles asked an identical question, that is, whether adipocytes would produce and release proteins that other cells did not make, while the fourth report approached the discovery from a different perspective. The first two articles actually used an almost overlapping experimental plan, exploiting the fact that adipocytes go through a maturation process before becoming specialized fat-storing cells. Thus, both teams examined adipocytes *in vitro* before and after these cells had matured and looked at which specific types of RNA (and consequently the proteins derived from them) were elevated in mature compared to immature adipocytes. This would be an indication that the proteins whose levels would be elevated after cell maturation were only produced by specialized adipocytes but not by cells that were still in an uncommitted state. Both groups ended up finding a specific type of RNA increasing more than a hundredfold as adipocytes matured, and both identified the protein encoded by this RNA, giving it two separate names (as usual), Acrp30 and AdipoQ. Though they both found a gene and its corresponding protein, neither group had a clue about this molecule's duties in the body. "We do not yet know the function of Acrp30," stated the Scherer report; "Its biological role remains to be defined," chimed in Dr. Hu's article. The authors of the third publication found adiponectin by asking a similar question as the first two papers but using a slightly different experimental strategy. Instead of looking at adipocytes before and after maturation, the investigators obtained small snippets of human tissues from various locations (fat, muscle, lung, liver, etc.) and examined which types of RNA showed up in adipose tissue but nowhere else. If they found any type of RNA that was only generated in adipose tissue, this would indicate the presence of proteins produced only by adipocytes, since these cells were for the most part absent in each of the other locations examined (a situation similar to the discovery that occasioned Jeff Friedman's eureka moment mentioned above). This search ended up with the very same RNA as the two previous papers, which in this third iteration was named apM1. Not surprisingly, the search also led to the same conclusion: the role of apM1 was not clear. Lastly, the fourth group of researchers was looking for something quite different from the first three teams. At variance with the other investigators, Dr. Nakano's group was not interested in proteins made by adipocytes but rather wanted to find out if there was any protein in the bloodstream that would physically interact with the extracellular matrix, the convoluted tridimensional scaffolding that surrounds and supports cells in every tissue. Based on prior evidence showing that some proteins present in the bloodstream could associate with specific components of the extracellular matrix, these investigators reasoned that blood must also contain proteins that interacted with other kinds of matrix components. It was actually sort of a wild guess, but it bore fruit in the form of adiponectin, which the researchers choose to name GBP28. By this time, however, the other three adiponectin papers had been published and therefore the discoverers of GBP28 were able to determine quickly that their protein was made by adipocytes and was identical to the one found by the other three teams. Despite the very different kind of chase they had been engaged in, the conclusion of the authors of the fourth paper should by now be familiar: the biological function of the newly discovered protein remained unknown.

Catching Up

Since no relevant prior evidence was available, the scientists who discovered adiponectin had to start from scratch to understand its function, a situation comparable to the one faced by the researchers who had identified IL-36, IL-37, and IL-38, the IL-1 family members discovered through a structure-first strategy. However, each of the four papers that had reported the discovery of adiponectin contained a few hints that could be used as stepping-stones to build further investigations. Thus, one article showed that insulin increased adiponectin's production by adipocytes, suggesting a link with glucose metabolism. Others pointed to how similar the molecular structure of adiponectin was to that of a certain family of proteins, perhaps a clue to a structure-function relationship. Importantly, and quite puzzling for a protein coming from adipocytes, Dr. Hu's article showed that adiponectin was counterintuitively lower in obese compared to lean rodents and humans. This was the exact opposite of what investigators had just demonstrated for leptin, where more fat meant more leptin. Not so for adiponectin, which was inexplicably reduced, rather than increased, in the obese, exactly when more of the cells that produced this adipokine were present in the body.

These were all interesting clues, but quite limited. What might all of this mean? What to do with this information? Answers were not easy to come by and it took three long years after the original four articles—until 1999—before adiponectin resurfaced in the scientific literature. The parallel with leptin, which three years after its identification was already being injected into obese people in an attempt to curb their appetite, is striking, to say the least.

The first article published after the initial group of four confirmed and extended the original observation that adiponectin was reduced in obese individuals. Consequential for the studies that immediately followed was the technical description, in the same paper, of an assay for rapid quantification of adiponectin in human samples, which would allow the article's authors and other scientists to begin comparing levels of adiponectin under a variety of circumstances.[38] With this assay in hand, a researcher could, for example, examine how blood concentrations of adiponectin changed with age, sex, or lifestyle, while another group could compare adiponectin levels in individuals affected or not affected by various diseases, or even one person before and after having received a drug or other intervention. Meanwhile, biochemists continued to examine adiponectin's complex structure and, despite enormous technical difficulties, managed to engineer and produce it in recombinant form.

Between 1999 and 2001, a combination of epidemiological investigations, *in vivo* studies quantifying adiponectin under various conditions and *in vitro* approaches using the recombinant protein, began to paint a collective picture that saw adiponectin deeply involved in the pathophysiology of cardiovascular disease and type II diabetes. Researchers also started to report that adiponectin exerted several important anti-inflammatory effects, showing that this adipokine could suppress synthesis of TNF by macrophages, inhibit pro-inflammatory signaling pathways, and reduce the amount of adhesion molecules (the chemical Velcro) on the surface of endothelial cells. Although these studies pointed to the possibility that adiponectin might exert beneficial effects in cardiovascular disease and other conditions, scientists were exceedingly careful in their conclusions. The concept of an adipocyte-derived molecule that would be protective against the typical comorbidities of obesity may in fact have seemed a little unreasonable at the time, when fat was customarily considered as a troublemaker, not a caring agent that would send out nurturing messages to the rest of the body.[39]

Then, in August 2001, two articles in the same issue of *Nature Medicine* strongly contributed to change the notion of adipose tissue as a solely detrimental actor by demonstrating that adipocyte-produced adiponectin was instead conducive to health in experimental models of type II diabetes and related conditions. The authors of these two reports used addition approaches to establish that administration of recombinant adiponectin ameliorated diabetes in animals, both groups independently arriving at the conclusion that insufficient amounts of adiponectin were implicated in the metabolic alterations that characterize type II diabetes.[40]

Additional epidemiological evidence connecting adiponectin to type II diabetes and cardiovascular disease soon emerged, together with studies implicating molecular pathways related to diabetes in the regulation of adiponectin synthesis. By the end of 2002, the concept that adiponectin coordinates a complex response that protects the body against development of type II diabetes and cardiovascular disease, and that these conditions are in turn associated with meager levels of adiponectin, was firmly established. Generation of adiponectin knockout and transgenic mice, identification of adiponectin receptors and their downstream signaling pathways, *in vitro* and *in vivo* studies using recombinant adiponectin, more evidence about the regulation of adiponectin synthesis, more clinical studies, more epidemiological studies: the work of many researchers, with some of the original investigators continuing to play a central role, and the development of ever more sophisticated investigational tools contributed to further delineate multiple aspects of adiponectin's biology and its involvement in health and disease.[41]

By now, accumulated evidence showing the presence of inadequate quantities of adiponectin in persons with metabolic diseases is exhaustive and completely convincing. Several studies have also demonstrated that high levels of adiponectin predict reduced risk of heart attacks, in a mirror image of CRP.[42] A question comparable to the one that was asked for CRP can thus be posed here: Does low adiponectin per se cause metabolic disease? As we have seen in the previous chapter, for CRP the answer to the reverse of this question has been a convincing no, with gene variants studies showing that this acute-phase protein does not sit in the pathogenetic pathway of cardiovascular disease. For adiponectin, that answer is less conclusive. Many scientists went combing human genetic diversity to check whether gene variants that reduce levels of adiponectin might alter risk of type II diabetes or cardiovascular disease, the same strategy used, in the opposite direction, for CRP. When examined separately, several of these investigations suggested a causative role for adiponectin, but when multiple studies were analyzed together in what is called a meta-analysis evidence did not always hold up.[43]

It may well be, then, that making only a little adiponectin is not, by itself, what causes type II diabetes or cardiovascular disease. However, this does not mean that examining adiponectin in these conditions is meaningless. Many excellent studies demonstrated that either artificially raising adiponectin levels or activating its receptors had beneficial effects in experimental models of metabolic diseases, regardless of what the ultimate cause of the disorder might be—if an ultimate cause can, in fact, be conceivable for this type of disorders. A parallel to the role of TNF in rheumatoid arthritis and inflammatory bowel disease seems appropriate here. As discussed in previous chapters, although TNF is not the fundamental cause of these chronic inflammatory

diseases, blocking this cytokine's activity can be a very effective therapeutic approach. Adiponectin may thus fall into a similar paradigm,
where one could obtain favorable effects by activating its receptors or
increase its concentration irrespective of whether adiponectin should
be placed in the primary causal path of metabolic disease.[44]

It took a while, but adiponectin eventually caught up with leptin.
From only seven articles containing original findings about adiponectin
in 1999 compared to the more than seven hundred about leptin, the
number of scientific reports on the two adipokines slowly converged,
with adiponectin climbing up to leptin. As had been for IL-1 family
members IL-36, IL-37, and IL-38, arriving into the world of body messages without a proposed function stalled progress. In the case of adiponectin, however, the initial obstacles were bridged more rapidly
than for the cytokines discovered through a structure-first strategy.
Some of the characteristics that were already listed in the four original
publications on adiponectin—its association with obesity and metabolism, its copious production—likely represented a sufficient motivation to draw investigators to the study of this adipokine and keep them
interested. The relatively swift accumulation of evidence of an association between adiponectin and the ever more common metabolic disorders, coupled with technical simplicity in measuring this adipokine,
attracted new minds, resulting in a critical mass that pushed the field
forward. However, specifically delineating the precise mechanisms by
which adiponectin promotes a healthy metabolism has been rather difficult, such that the picture of adiponectin's biology remains considerably blurry.

Intricate Connections

There are many reasons for that blur. Adiponectin's structure, with its
ability to assemble into supramolecular complexes but also generate
small fragments, has complicated understanding of whether the various forms this adipokine can assume might exert differential effects.
The process of adiponectin synthesis and release from adipocytes has
also been enigmatic. Even finding its receptors took considerably
longer than one would expect, as eight years passed between the discovery of adiponectin and cloning of its two receptors, where it took

only one year for leptin. It didn't help that adiponectin receptors are so peculiar that they do not belong to any known family and that, even after definition of their detailed atomic structure by crystallography, how adiponectin attaches to its receptors remains undefined. There may even be a third receptor, but not everyone agrees. If those issues were not enough, one should also consider that adiponectin sits at the juncture of several key nodes of the intricate network that coordinates both metabolic and inflammatory responses, making it almost impossible to extricate the tangle by analyzing the various pathways separately. Alter one part and the others are also inevitably affected.[45]

One can begin by admiring the reciprocal dance adiponectin and leptin are constantly engaged in. Leptin levels rise during feeding or in response to excess calories, when adipocytes expand in both size and number. In contrast, adiponectin levels increase when calories are insufficient or physical activity is elevated, adipocytes thus reducing their size as triglycerides are used to generate energy. "The 'yin and yang' of these two adipokines, which are oppositely regulated under essentially all conditions, suggests coordinate regulation," write Drs. Roger Unger, Philipp Scherer, and William Holland in an engaging 2013 review article.[46] In obese individuals, leptin is up, adiponectin down, and adipocytes are large and abundant. In contrast, in people who have only little adipose tissue (because they cannot eat a lot, choose not to do so, or undergo intense physical activity), leptin is low, adiponectin high, and adipocytes are small and sparse. A comparable situation—on a more limited scale—is repeated during the feeding-fasting cycle of everyday life (at least in mice), where leptin reaches its zenith after meals and adiponectin instead does it during the prolonged fasting of sleep. If leptin has been dubbed the satiety factor, adiponectin has instead been conceived of as the starvation hormone.[47]

These intricate connections make it almost impossible to disentangle the two adipokines from their very intimate dance, to understand the function of one independently of the other. Take *ob* mice for example, the mutants that do not produce leptin. These mice are massively obese, as we have seen, because they miss the hunger brake of leptin. However, their ample and copious adipocytes and constant state of overnutrition also mean that their adiponectin levels are low. Therefore, *ob* mice are not only leptin-deficient, they are also adiponectin-low, and understanding what's responsible for what becomes extremely difficult. Only the experimental tricks of shrewd investigators can disentangle the two dancers. Dr. Scherer's team did just that by genetically engineering *ob* mice to produce plentiful adiponectin while leaving the animals without leptin. This tweaking resulted in remarkable changes. Mice with no leptin but a lot of adiponectin became even fatter than those that had been left leptin-deficient and adiponectin-low. But, despite weighing almost two times their siblings, the engineered mice were healthier. Their sugar and lipid metabolism improved, their liver looked so much better. The mice had become fat and fit, at least from a metabolic perspective.

What happened? Let's recall an observation we mentioned at the beginning of this chapter: one of the proposed evolutionary roles of adipocytes is to protect other cells from the toxic effects of fat. Having more adipocytes, therefore, may actually not be that bad, since a larger amount of fat can be sequestered away inside adipocytes and therefore protect other tissues from the potential toxicity of accumulated lipids, which is just what seemed to have happened in Dr. Scherer's genetically twisted mutants. One of the ways by which adiponectin can benefit metabolism is therefore by promoting storage of fat in its proper place—the adipocyte—instead of letting lipids build up in places where they would wreak havoc, such as the heart, muscles, liver, or pancreas.[48]

This may sound somewhat confusing: Is having a lot of adipose tissue a good or a bad thing? Didn't we just say a few pages back that obesity increases risk of several diseases? It turns out that the type and location of adipose tissue may count more than its absolute amount. Lots of large adipocytes, especially if they sit in the belly area, disrupt the harmonious balance of metabolic pathways, whereas adipose tissue made of small adipocytes positioned under the skin, particularly if they are in the bottom part of the body, seems to protect the heart, liver, and other organs from lipid toxicity. Better to be a pear than an apple. The mice engineered by the Scherer group experienced just such a change in adipose tissue composition. Even though mice that had been manipulated to be leptin-deficient and adiponectin-high were really really fat, their adipocytes did not have the mammoth size typical of *ob* mice, instead remaining small and thus exerting a protective function against the ravages of lipid toxicity.[49]

Though definitely intricate, a dance between two fat-derived hormones seems like a solvable problem. There is, however, an additional twist to the story. At the end of the previous chapter, I mentioned that the giant adipocytes of obese individuals dispatch SOS signals to the immune system, which responds with development of inflammation that, in turn, regulates risk and severity of obesity's comorbidities. We also discussed the ambiguous role of IL-6 in obesity-associated type II diabetes.

The fact that Scherer's leptin-deficient, adiponectin-high mice had small adipocytes means that their adipose tissue was not only better at storing fat, but also appreciably less inflamed than that of their siblings with large adipocytes, producing lower amounts of inflammatory mediators, including much less IL-6. Could lower inflammation have contributed to the better health status of the engineered mice? Possibly. But this observation brings us all the way back to the uncertainties about the role of inflammatory mediators—particularly IL-6—in modulating metabolic pathways.

We learned in the preceding chapter that the rise in IL-6 that accompanies obesity might actually represent a protective response against disruption of sugar metabolism. So, was the low IL-6 of leptin-deficient, adiponectin-high mice contributing to making these animals healthier or was it instead opposing the beneficial effects of adiponectin? Answering this question becomes even more problematic if we consider that adiponectin may actually need IL-6 itself to protect against type II diabetes. It's truly a morass of intertwined parts, many more than I can examine here, and I will therefore leave it to Drs. Unger, Scherer, and Holland: "Over the past two decades, the discoveries of adipocyte-derived factors ('adipokines') and their mechanistic actions have left us marveling at and struggling to understand the role these factors serve in physiology and the pathophysiology of obesity and diabetes."[50]

Although scientists are still struggling to disentangle these pathways, we can nevertheless attempt to combine all this evidence into a general picture. An excess of calories accompanied by a sedentary lifestyle leads to development of inflamed adipose tissue made of large adipocytes that produce very little adiponectin. The resulting low adiponectin fails to protect the organism against the damages of lipid toxicity while being unable to regulate sugar metabolism and suppress inflammation. As a consequence, low adiponectin is associated with disrupted metabolism and increased inflammation, amplifying the risk of type II diabetes, cardiovascular disease, and other disorders. Inflammation and disrupted metabolism then feed back into the system, further curtailing adiponectin and perpetuating the detrimental cycle.

In contrast, maintaining a healthy weight through proper diet and physical activity allows adipocytes to remain small and nimble, preventing development of adipose tissue inflammation. In turn, these small adipocytes squeeze out lots of adiponectin, which works on the liver, muscles, and other tissues and organs to maintain a balanced metabolism and an anti-inflammatory tone. Here, too, the system feeds back into itself, favoring elevated adiponectin and thus perpetuating the healthful cycle. But one needs to actively maintain this cycle of health, so that sufficient adiponectin keeps being generated.

In 1930, German physician-scientist-medical anthropologist Viktor von Weizsäcker provided a definition of health that remains my favorite of all times, "Health is not a capital resource which one may exhaust, but exists only so long as it is continually generated. If it ceases to be generated, one is already ill." I sincerely doubt anybody could come up with a more adequate description for adiponectin.[51]

A Closing Rally

In discussing how adipokines revolutionized our understanding of adipose tissue biology, in this last chapter we rallied and reiterated several of the book's major themes. The history of the discovery of leptin and adiponectin allowed us to review the challenges and advantages of using a function-first versus a structure-first approach. We could also appreciate the convoluted roads that inextricably commingle clinical and basic research, while acknowledging that both *in vivo* and *in vitro* observations can act as the initial spark for new and highly productive lines of investigation. We saw that observation and intervention studies, addition and subtraction approaches, sheer determination and random luck each play valuable and distinct roles in the process of discovery, and pondered how nonlinear the research process can be. Finally, we reintroduced the futility and practical impossibility of neatly separating areas of investigation that have been defined as if our bodies were compartmentalized assemblages of parts, not whole organisms without borders.

Epilogue

IF WE DECIDE to agree with von Weizsäcker that health is a resource that needs to be continually generated, we may also find consensus about the importance of doing so from a biological body that is undivided by cultural constructs.

In the last chapter, we pondered about how ensnared adiponectin is amid metabolism and inflammation, so much so that its precise location in the world of biomedical research cannot be precisely pinned down. However, the whereabouts of the other body messages we got to know in these pages are no less ambiguous.

Chased for decades as the molecular basis for the lipostat theory, leptin—the other adipokine—is so much more than a satiety factor. Following up on early reports about infertility, altered sympathetic nervous system activity, fragile bones, and defective immunity in *ob* mice that lack leptin and *db* mice that lack receptors for this adipokine, studies eventually demonstrated that leptin is an inherent component of the regulation of fertility while also participating in the control of blood pressure, bone metabolism, and immune responses. The increased susceptibility of malnourished individuals to infections, for instance, is in part attributable to the dramatic reduction in leptin levels secondary to depletion of fat reserves, this adipokine acting as a regulator of lymphocytes survival and proper functioning. Infections and their associated inflammation, in turn, push adipocytes to produce more leptin, possibly contributing to modulating the response against microbial pathogens. Thus, just like adiponectin, the study of leptin is also split among more than one disciplinary specialty.[1]

And, by all means, adipokines are not alone in their belonging to multiple worlds. Hepcidin, the iron regulator presented in Chapter 6, straddles the fields of infection, inflammation, metabolism, endocrinology, and nutrition in a manner not dissimilar from that of leptin and adiponectin, even if via altogether different mechanisms. In Chapter 6 we also reviewed the involvement of IL-6 in regulation of glucose homeostasis, discussing how this cytokine sits at the intersection between inflammation and metabolism, while many additional functions of IL-6 put this mediator into a hard-to-classify position. Indeed, each of the other body messages introduced in the previous chapters, as well as many of those we did not get to meet, face a similar karma.

Scientists initially identified TNF, in its incarnation as cachectin, based on this cytokine's ability to induce wasting by altering lipid metabolism. However, even though TNF's metabolic activities continued to be studied, research soon overwhelmingly focused on the involvement of this body message in inflammatory responses and cell death, as we saw in Chapters 4 and 5. Beginning in the mid-1990s, however, increased rates of obesity slowly brought the metabolic aspects of TNF center stage again. This time, however, investigators approached the study of this cytokine mostly for its role with having too much fat rather than too little, as had been the case when Dr. Cerami was puzzled by his observations in skinny cachectic cows. Over the years scientists' understanding of TNF therefore made a double flip: from a molecule that caused death of tumors to one that made cancer thrive (as reviewed in Chapter 4) and from a mediator of wasting to a major player in the pathologies that accompany obesity.[2]

Many members and handlers of the IL-1 family (presented in Chapter 3) have also lived a double life. Fever was one the routes that led researchers to identification of one of the family founders—IL-1β— as part of the inflammatory response to infections. But what is fever if not elevated and deregulated metabolism brought about by inflammatory mediators? We should also not forget the name catabolin, bestowed early on upon IL-1 to indicate this cytokine's ability to destroy joints by promoting catabolic responses that break apart large biomolecules.

Just as it has been for its brother-in-arms TNF, scientists have mostly investigated IL-1 in the context of inflammation and immunity, though an underground current of studies on IL-1's involvement in metabolic dysregulation has always been present. As a more encompassing understanding of the connection between metabolic diseases and inflammation arose in the last two decades, boosted by research on obesity, interest in IL-1 as an integral mediator of the pathogenesis of cardiovascular disease, diabetes, and other obesity comorbidities rapidly intensified, with extensive efforts to target IL-1 as a therapeutic strategy in disorders characterized by a prominent metabolic component. Several additional members of the IL-1 family, as well as many other cytokines, have been involved in the pathogenesis of various metabolic diseases.[3]

The IL-1 family handler caspase-1, which activates IL-1β and IL-18 (as discussed in Chapter 3), is no less a participant in the connection between inflammation and metabolism. Multiple studies demonstrated involvement of this enzyme and its activating apparatus, the inflammasome, in the inflammation that consorts with obesity and metabolic diseases. Then, in a meaningful twist of biology for a society afflicted by excess girth, the notion that one could starve inflammation ended up losing its metaphorical meaning when investigators discovered that β-hydroxybutyrate, a by-product of fat metabolism produced by the liver during fasting, inactivates the inflammasome, thus suppressing inflammation.[4]

The list could continue, but the concept at this point should be clear: metabolism profoundly affects inflammation and immunity, while inflammatory and immune responses are inextricably linked with metabolism, the separation of these fields becoming ever more tenuous and unnecessary, perhaps even detrimental to progress. This interlacing is hardly limited to the few areas covered in these pages, rather extending to every field of biomedical research, like a self-generating filigree where each thread is connected to all other threads.

If we thus concur with the concept that the biomedical research of the future needs a biology that is whole, carried out by investigators who understand the importance of blending disparate disciplines and approaches into a common language that appreciates the organism as a world that is truly without borders, what better fit than the science of body messages?

Abbreviations

AIDS	acquired immuno deficiency syndrome
Anxa1	annexin A1
CRP	C-reactive protein
DNA	deoxyribonucleic acid
HDL	high-density lipoprotein
HIV	human immunodeficiency virus
Ig	immunoglobulin
IGIF	interferon-γ-inducing factor
IL	interleukin
IL-18BP	interleukin-18 binding protein
IL-1Ra	interleukin-1 receptor antagonist
IL-36Ra	interleukin-36 receptor antagonist
IBD	inflammatory bowel disease
JAK	Janus-activated kinase
MCP-1	monocyte chemotactic protein-1
MODY	maturity-onset diabetes of the young
NF-κB	nuclear factor-κB
NLRP3	NACHT, LRR and PYD domain-containing 3
PKC	protein kinase C

PRR	pattern recognition receptor
RNA	ribonucleic acid
SAA	serum amyloid A
SCID	severe combined immunodeficiency
SOCS	suppressor of cytokine signaling
STAT	signal transducer and activator of transcription
TBP	tumor necrosis factor binding protein
Th	T helper
TNF	tumor necrosis factor

Notes

FOREWORD

1. W. M. Bayliss and E. H. Starling, "Croonian Lecture: The Chemical Regulation of the Secretory Process," *Proc R Soc Lond* 73 (1904): 310–322.
2. Ibid., 315.
3. Stephen Kern, *The Culture of Time and Space, 1880–1918* (Cambridge, MA: Harvard University Press, 2003).
4. J. Henderson, "Ernest Starling and 'Hormones': An Historical Commentary," *J Endocrinol* 184 (2005): 5–10.
5. E. H. Starling, "Croonian Lecture: The Chemical Correlation of the Functions of the Body," *Lancet* 4275 (1905): 339–341, quote at 339.
6. E. H. Starling, "Croonian Lecture: The Chemical Correlation of the Functions of the Body, Lecture II," *Lancet* 4276 (1905): 423–425, quote at 423.
7. E. H. Starling, "Croonian Lecture: The Chemical Correlation of the Functions of the Body," *Lancet* 4275 (1905): 339–341, quote at 339.

CHAPTER 1   The Medium Is the Message, the Message Is the Change

1. Marshall McLuhan, *Understanding Media: The Extensions of Man* (New York: Signet Books, 1964). "The Medium Is the Message" is the title of the book's first chapter.
2. Body messages can also take unexpected and complex forms, such as the versatile, minuscule microRNAs (and other forms of small nucleic acids) that can convey information through the systemic circulation and other body fluids or the supramolecular structures of microvesicles and exosomes, with the potential of transmitting multifaceted, composite information from one location to another. For more on these forms of body messages, see, for

example, Matthew Weiland et al., "Small RNAs Have a Large Impact: Circu-lating MicroRNAs as Biomarkers for Human Diseases," *RNA Biology* 9 (2012): 850–859; Dylan Burger et al., "Microparticles: Biomarkers and Beyond," *Clin Sci* 124 (2013): 423–441.

3. One of the main genetic sources of variation among humans are the genes of the human leukocyte antigen system, also called major histocompatibility complex. For a story of the discovery and biology of this complex system, see the engaging book by Daniel M. Davis, *The Compatibility Gene* (London: Allen Lane, 2013). It is important to remember that the environment and our life experiences are potent modifiers of the way DNA operates (although not of its sequence) through mechanisms known as epigenetics. Genes only con-tain information for proteins: therefore, there is no gene for nitric oxide, estrogen, testosterone, cortisol, or prostaglandins, which are not proteins; there is no DNA sequence that codes for these and all the other body mes-sages that are not proteins. Hence, unlike the characters of this book, these nonprotein body messages have the exact same structure in each human being. However, the enzymes necessary to make these messages are proteins and thus genetic variability indirectly affects the way in which even nonpro-tein messages are produced.

4. Some body messages work both outside and inside the cell of origin.

5. Some of the smallest molecular messages, like nitric oxide, do not have spe-cific receptors and act by directly modifying their molecular targets.

6. Some types of receptors are instead present inside receiver cells and work a little bit differently than described here, though the general concept that binding of the ligand activates the receptor also applies to them.

7. Queried on January 17, 2014.

8. McLuhan, *Understanding Media*, 10.

9. Kenneth E. Boulding, *The Image: Knowledge in Life and Society* (Ann Arbor: University of Michigan Press 1956), 7.

10. Several other investigators participated in the discovery and characterization of PRR, among whom Dr. Ruslan Medzhitov played a prominent role. Dr. Ralph Steinman was the other awardee of the 2011 Nobel Prize in Physiology or Medicine for his contribution to the discovery of a type of immune cells called dendritic cells. Moreover, the discovery of toll, a fruit fly gene, was instrumental in fostering the field of PRR studies. In 1995, the discovery of toll contributed to selection of German scientist Dr. Christiane Nüsslein-Volhard as the recipient of the Nobel Prize in Physiology or Medicine, which she shared with Drs. Edward B. Lewis and Eric F. Wieschaus.

11. There are other types of leukocytes in addition to monocytes/macrophages, neutrophils and lymphocytes, including dendritic cells, mast cells, natural killer cells, eosinophils and basophils.

12. For more on current interpretations of the immune system, see Bastiaan A. Blok et al., "Trained Innate Immunity as Underlying Mechanism for the Long-Term, Nonspecific Effects of Vaccines," *J Leukoc Biol* 98 (2015): 347–356; Akiko Iwasaki and Ruslan Medzhitov, "Control of Adaptive Immunity by the Innate Immune System," *Nat Immunol* 16 (2015): 343–353; David

Artis and Hergen Spits, "The Biology of Innate Lymphoid Cells," *Nature* 517 (2015): 293–301.

13. Hannah Landecker, "Postindustrial Metabolism: Fat Knowledge," *Public Culture* 25 (2013): 495–522; Gordon M. Tomkins, "The Metabolic Code," *Science* 189 (1975): 760–763.

14. Landecker, "Postindustrial." The same definition could be used to define the immune response as a whole, not just inflammation.

15. For example, see Gökhan S. Hotamisligil and Ebru Erbay, "Nutrient Sensing and Inflammation in Metabolic Disease," *Nat Rev Immunol* 12 (2008): 923–934; Raj Chovatiya and Ruslan Medzhitov, "Stress, Inflammation, and Defense of Homeostasis," *Mol Cell* 54 (2014): 281–288; Abiskek Iyer et al., "Nutrient and Immune Sensing Are Obligate Pathways in Metabolism, Immunity, and Disease," *FASEB J* 29 (2015): 3612–3625.

CHAPTER 2 Doing Research: Ways to Study Body Messages

1. The World Health Organization defines epidemiology as the study of the distribution and determinants of health-related states or events (including disease), and the application of this study to the control of diseases and other health problems.

2. Possible criteria include age, sex, ethnicity, economic and educational status, whether people smoke or drink alcohol, whether they take certain drugs or supplements, how much they exercise, where they live, and many more.

3. The decision about the number of subjects to include in a study is essential because it determines the statistical power of the investigation. However, although it should be mostly based on statistical power, the size of the groups to be included in a study needs to be counterbalanced by issues of feasibility and cost.

4. The word *passive* to describe observational studies is not meant to carry any negative connotation but is rather used to conceptually oppose this type of research to the purposefully active approach of interventional investigations.

5. As most readers are probably aware, the use of experimental animals in biomedical research is a highly controversial issue. I will briefly tackle the ethics of animal use in research later in this chapter as well as in other places in the upcoming pages.

6. It is important to stress that any procedure on a human or animal inevitably carries some sort of risk. However, the risk of obtaining cells from humans is usually negligible compared to a direct intervention in the whole organism as required by *in vivo* experiments.

7. Rebecca Skloot, *The Immortal Life of Henrietta Lacks* (London: Macmillan, 2010).

8. Regulatory steps are necessary to obtain permission to buy cells. Usually, cell lines and other laboratory materials are only available to individuals who are part of an organization involved in biomedical research.

9. Readers interested in the history of cell culture and how it changed our

concept of life should read Hannah Landecker, *Culturing Life: How Cells Became Technologies* (Cambridge, MA: Harvard University Press, 2007).

10. Informed consent is at the basis of ethical conduct of research in human subjects. There are exemptions to the rule of informed consent for samples that have already been collected for past studies, provided the individuals involved in those studies had given approval to additional analyses. Each country has its own requirements for participation in research involving humans. For more information on international regulations and agreements on ethical principles for medical research involving human subjects, see the Declaration of Helsinki at http://www.wma.net/en/30publications/10policies /b3/index.html.

11. Experts in research ethics have written many excellent articles and books to which the reader is referred for a comprehensive review of this topic. See, for example, Marion Danis, Benjamin Berkman, and Sara Chandros Hull, *Research Ethics Consultation: A Casebook* (Cary, NC: Oxford University Press, 2012), and Robert Hubrecht, *Animals in Research: An Introduction to Welfare and Ethics* (Somerset, NJ: John Wiley, 2014). Ethical issues also apply to *in vitro* and *ex vivo* studies involving both humans and animals. In the chapters that follow, I will present the results of scientific studies involving both experimental animals and human subjects. A part of these studies, particularly some of those performed in animals before strict ethical standards had been put in place, may sound unacceptable to contemporary readers. Inclusion of the description of these experiments does not constitute an endorsement on my part of the activities performed, though I should clarify that I have employed rodents as experimental animals in my own research.

12. Unfortunately, several of these genetic conditions are so rare, at times affecting just a handful of people worldwide, that pharmaceutical companies often do not have sufficient financial incentives to develop drugs specifically targeted to treatment of such rare conditions, even when the underlying cause of the disease has been discovered. However, as we will see, the study of body messages nevertheless resulted in treatments for several types of rare genetic diseases, usually as an aftereffect of development of medications adressed to much more common conditions; this markedly improved life of at least some individuals affected by rare genetic diseases, and at times literally made the difference between life and death.

13. With perfection of ever newer tools for biological research, particularly genetic manipulation of cells and organisms, it is now possible to implement each of the two approaches discussed in this section—addition and subtraction—in sophisticated ways, using transgenic animals or cells that produce high amounts of a given protein or their opposite, knockout organisms that do not have the gene for a certain protein and are therefore unable to make a given mediator or its receptor. These systems can be extremely complex and precise, adding or removing production of one or more proteins under strictly controlled circumstances. However, with sophistication comes the need to be exceptionally careful in planning and interpreting results, to be aware of the limitations and quirks that accompany each technique. Data obtained

using high-tech methodologies are not always necessarily the strongest, even though they can be very powerful.

CHAPTER 3   Evolution of Message Discovery: The Interleukin-1 Family

1. Charles A. Dinarello, "Overview of the Interleukin-1 Family of Ligands and Receptors," *Semin Immunol* 25 (2013): 389–393.

2. The quote is from Charles A. Dinarello, "Celebrating 25 Years. IL-1: Discoveries, Controversies and Future Directions," *Eur J Immunol* 40 (2010): 595–653. For full disclosure, Dr. Dinarello has been my post-doctoral mentor and we remained good friends after I left his research group to start my own.

3. William Osler, "The Study of the Fevers of the South," *JAMA* 26 (1896): 999–1004. For a history of fever in its various incarnations, see Christopher Hamlin, *More than Hot. A Short History of Fever* (Baltimore: Johns Hopkins University Press, 2014). Dr. Bartfai's quote is from a January 2014 e-mail interview.

4. Elisha Atkins and W. Barry Wood, Jr., "Studies on the Pathogenesis of Fever. I. The Presence of Transferable Pyrogen in the Blood Stream Following the Injection of Typhoid Vaccine," *J Exp Med* 101 (1955): 519–528. For references to many of the original articles on fever research, see Dinarello, "Celebrating 25 Years."

5. William M. Ord, "Address on Some of the Conditions Included Under the General Term 'Rheumatoid Arthritis,'" *Brit Med J* 1 (1880): 155–158. At the time, the term *rheumatic* was rather imprecise but was beginning to signify inflammation secondary to a bacterial infection, while *scrofulous* refers to a type of tuberculosis.

6. For original references, see Jean-Michel Dayer, "The Saga of the Discovery of IL-1 and TNF and Their Specific Inhibitors in the Pathogenesis and Treatment of Rheumatoid Arthritis," *Joint Bone Spine* 69 (2002): 123–132.

7. Recall that *in vitro* proliferation of T lymphocytes was one of the three paths leading to identification of IL-1. Several investigators walked this *in vitro* road, including Drs. Fritz H. Bach, Igal Gery, Kouji Matsushima, Steven Mizel, Joost Oppenheim, Byron Waksman, and others. Although I am using the term IL-1 here, this terminology had not yet been introduced at the time of these studies.

8. Dr. Dinarello's quote and subsequent statements are from an August 2014 interview. Nowadays, fever in experimental animals is measured with radio thermometers. See Tamas Bartfai and Bruno Conti, "Fever," *TheScientificWorldJournal* 10 (2010): 490–504.

9. The NIH Associate Training Program, a section of US legislation commonly known as "Doctor Draft," allowed young male physicians, between 1965 and 1973, to fulfill their military obligation doing clinical and research work at the NIH. Though the program was rightly seen as highly discriminatory, since it allowed only those already privileged enough to have obtained a medical degree to escape the horrors and dangers of war, it ended up being one of the most important sources of physician-scientists at the time, spawning nine Nobel Laureates, ten recipients of the National Medal of Science,

sixty-four members of the National Academy of Sciences (including Dr. Dinarello), and a host of distinguished researchers, some of whom made critical contributions to the science of body messages. See Sandeep Khot, Buhm Soon Park, and William T. Longstreth, Jr., "The Vietnam War and Medical Research: Untold Legacy of the U.S. Doctor Draft and the NIH Yellow Berets," *Acad Med* 86 (2011): 502–508.

10. The name IL-1 was initially assigned to the message that had been discovered following the path of lymphocyte proliferation. Only a few months later was a study demonstrating the identity of this factor with leukocytic pyrogen published. References to the original reports are in Dinarello "Celebrating 25 Years" and Dayer "The Saga.".

11. Catabolism, which in Greek means throwing down, is the part of metabolism that breaks down the chemical bonds of large molecules to release energy and/or recycle the resulting small individual molecules that are liberated into new large structures. The arm of metabolism that instead builds things up is known as anabolism (throwing upward).

12. Fever itself can be viewed as a metabolic process, a speeding up of biochemical reactions brought about by an increase in body temperature. Prolonged fever activates catabolic pathways, so there is a close link between IL-1, inflammation, and metabolism whether one looks at fever, joint damage, or some of the other activities of IL-1 not described here.

13. This refers to a common laboratory technique.

14. Philip E. Auron et al., "Nucleotide Sequence of Human Monocyte Interleukin 1 Precursor cDNA," *PNAS* 81 (1984): 7907–7911. For a first-person history of the purification and cloning of IL-1, see Charles A. Dinarello, "The History of Fever, Leukocytic Pyrogen and Interleukin-1," *Temperature* 2 (2014): 8–16.

15. Peter T. Lomedico et al., "Cloning and Expression of Murine Interleukin-1 cDNA in Escherichia coli," *Nature* 312 (1984): 458–462. As mentioned in note 7, the *in vitro* lymphocyte proliferation route to IL-1 was followed by many investigators, mostly immunologists interested in the immune-modulating, rather than the pro-inflammatory, functions of IL-1. The fact that I did not explain this route in more detail has everything to do with space issues and nothing to do with its importance relative to the other two paths.

16. Carl J. March et al., "Cloning, Sequence and Expression of Two Distinct Human Interleukin-1 Complementary DNAs," *Nature* 315 (1985): 641–647.

17. For more details on the discovery of the IL-1 receptor(s), see Chapter 5.

18. Sheldon M. Wolff et al., "Clone Controversy at Immunex," *Nature* 319 (1986): 270; Auron et al.,"Nucleotide Sequence."

19. March et al., "Cloning." For a witness account of the events at Schloss Elmau, see Lawrence B. Lachman, "Summary of the Fourth International Lymphokine Workshop," *Lymphokine Res* 4 (1985): 51–57.

20. Quotes are from my interview with Dr. Dinarello. For more details on this case, see Dinarello, "The History"; Eliot Marshall, "Suit Alleges Misuse of Peer Review," *Science* 270 (1995): 1912–1914; Eliot Marshall, "Trial Set to Focus on Peer Review," *Science* 273 (1996): 1162–1164; Eliot Marshall, "Battle Ends in $21 Million Settlements," *Science* 274 (1996): 911. Immunex

scientists eventually did clone IL-1β and that's how they corrected the nine mistakes present in their patent application, reporting the correct sequence in their June 1985 *Nature* article.

21. Giorgio Prodi, "Signs and Codes in Immunology (1988)," in *Essential Readings in Biosemiotics: Anthology and Commentary*, Biosemiotics 3, ed. Donald Favareau (Dordrecht: Springer Netherlands, 2009), 332.

22. Quotes and personal information from Dr. Dayer are from a May 2014 e-mail interview. For references to the original IL-1 articles, see Dayer, "The Saga."

23. Quotes and personal information from Dr. Arend are from a January 2014 e-mail interview.

24. These are clearly imprecise parallels, since the level of alcohol is someone's breath is also influenced by rates of alcohol absorption, body size, etc.

25. Roger Guillemin and Andrew V. Schally shared with Rosalyn Yalow the 1977 Nobel Prize in Physiology or Medicine for their discoveries concerning peptide hormone production. For an elaboration on the concept of old knowledge, see Helena Worthen, *What Did You Learn at Work Today? The Forbidden Lessons of Labor Education* (New York: Hard Ball Press, 2014), 175–199.

26. Charles A. Dinarello, Lanny J. Rosenwasser, and Sheldon M. Wolff, "Demonstration of a Circulating Suppressor Factor of Thymocyte Proliferation During Endotoxin Fever in Humans," *J Immunol* 127 (1981): 2527–2519; James W. Larrick, "Native Interleukin 1 Inhibitors," *Immunol Today* 10 (1989): 61–67.

27. William P. Arend, Fenneke G. Joslin, and Joseph Massoni, "Effects of Immune Complexes on Production by Human Monocytes of Interleukin 1 or an Interleukin 1 Inhibitor," *J Immunol* 134 (1985): 3868–3875. In 1985, Drs. Arend and Dayer both presented their findings at a conference in Ann Arbor, MI. Both articles are published in Matthew J. Kluger, Joost J. Oppenheim, and Michael C. Powanda, eds., *The Physiologic, Metabolic, and Immunologic Actions of Interleukin-1. Progress in Leukocyte Biology*, vol. 2 (New York: Alan R. Liss, 1985). The respective references are William P. Arend et al., "Interleukin 1 Production by Human Monocytes: Effects of Different Stimuli," 399–407, and Jean-François Balavoine et al., "Collagenase- and PGE_2-Stimulating Activity (Interleukin-1-Like) and Inhibitor in Urine from a Patient with Monocytic Leukemia," 429–436.

28. An early version of this work was presented in 1984 at the same conference at Schloss Elmau where the drama involving cloning of IL-1 we just reviewed had taken place: Jean-François Balavoine et al., "Identification of Interleukin 1-Like Activity and Inhibitor(s) in Urine from a Patient with Acute Monoblastic Leukemia," *Lymphokine Res* 3 (1984): 233.

29. Dr. Dayer's article demonstrating the receptor antagonist activity of the new inhibitor was published in the *Journal of Immunology* after having been sequentially rejected by *Nature* and *The Journal of Experimental Medicine*. "They did not recognize the importance of this new concept, the first example of a natural antagonist blocking the binding of a cytokine to its receptor," Dr. Dayer told me in our interview. That article is Philippe Seckinger et al., "A Urine Inhibitor of Interleukin 1 Activity That Blocks Ligand Binding,"

J Immunol 139 (1987): 1546–1549. Dr. Arend's article is William P. Arend et al., "An IL-1 Inhibitor from Human Monocytes. Production and Characterization of Biologic Properties," *J Immunol* 143 (1989): 1851–1858.

30. The article that introduced the name IL-1Ra is Charles H. Hannum et al., "Interleukin-1 Receptor Antagonist Activity of a Human Interleukin-1 Inhibitor," *Nature*, 343 (1990): 336–340. For references to articles that also explain some of the technical approaches used, see William P. Arend and Mary B. Goldring, "The Development of Anticytokine Therapeutics for Rheumatic Diseases," *Arthritis Rheumatism*, 58 (2008): S102–109; Jean-Michel Dayer, "Evidence for the Biological Modulation of IL-1 Activity: The Role of IL-1Ra," *Clin Exp Rheum* 20 (2002): S14–20. For additional information, see William P. Arend and Jean-Michel Dayer, "Cytokines and Cytokine Inhibitors or Antagonists in Rheumatoid Arthritis," *Arthritis Rheum* 33 (1990): 305–315.

31. All quotes and personal recollections are from my interviews with Drs. Arend and Dayer. For an additional cytokine-related story on how frustration in interpreting bioassays' results can lead to important discoveries, see Alfons Billiau, "A Tale of Two Interferon Bioassays: How Frustration with Discrepant Results from Slightly Dissimilar Methods Can Engender Discovery," *Method Mol Biol* 820 (2012): 1–6.

32. To learn more about therapeutically blocking IL-1, see Charles A. Dinarello and Jos W. M. van der Meer, "Treating Inflammation by Blocking Interleukin-1 in Humans," *Semin Immunol* 25 (2013): 469–484.

33. Quotes and personal information from Dr. Tsutsui are from a February 2014 e-mail interview.

34. For wonderful histories of interferon-γ's discovery, see Alfons Billiau and Patrick Matthys, "Interferon-γ: A Historical Perspective," *Cytokine Growth F R* 20 (2009): 97–113; Jan Vilcek, "My Fifty Years with Interferon," *J Interf Cytok Res* 27 (2007): 535–541. For the role of mutations in susceptibility to infections, see Laurent Able et al., "Human Genetics of Tuberculosis: A Long and Winding Road," *Philos T R Soc B* 369 (2014): 20130428.

35. Haruki Okamura et al., "Cloning of a New Cytokine That Induces IFN-γ Production by T Cells," *Nature* 378 (1995): 88–91. For more information on IL-12, see Giorgio Trinchieri, "Interleukin-12 and the Regulation of Innate Resistance and Adaptive Immunity," *Nat Rev Immunol* 3 (2013): 133–146.

36. For more on IL-18's biology, see Hiroko Tsutsui and Kenji Nakanishi, "Immunotherapeutic Applications of IL-18," *Immunotherapy* 4 (2012): 1883–1894; Daniela Novick et al., "Interleukin-18, More Than a Th1 Cytokine," *Semin Immunol* 25 (2013): 439–448.

37. J. Fernando Bazan, Jackie C. Timans, and Robert A. Kastelein, "A Newly Defined Interleukin-1?" *Nature* 379 (1996): 591.

38. For references to the original reports, see Eleanor Dunn et al., "Annotating Genes with Potential Roles in the Immune System: Six New Members of the IL-1 Family," *Trends Immunol* 22 (2001): 533–536.

39. See Martha Slud, "Boom Time for Biotech," *CNN Money*, February 22, 2000, http://money.cnn.com/2000/02/22/companies/biotech/.

40. For references to the original articles, see Cem Gabay and Jennifer E. Towne,

"Regulation and Function of Interleukin-36 Cytokines in Homeostasis and Pathological Conditions," *J Leukoc Biol* 97 (2015): 645–652.

41. Ibid.

42. Gabay and Towne, "Regulation"; Matthias Eberl, "Don't Count Your Interleukins Before They've Hatched," *Trends Immunol* 23 (2002): 341; Claudia A. Nold-Petry et al., "IL-37 Requires the Receptors IL-18Rα and IL-1R8 (SIGIRR) to Carry Out Its Multifaceted Anti-Inflammatory Program Upon Innate Signal Transduction," *Nat Immunol* 16 (2015): 354–365; Xianli Yuan et al., "Role of IL-38 and Its Related Cytokines in Inflammation," *Mediators Inflamm* 2015 (2015): 807976.

43. Dr. Severo Ochoa shared the 1959 Nobel Prize in Physiology or Medicine with Dr. Kornberg; Drs. Seymour Benzer and Sydney Brenner shared the 1971 Lasker Award with Dr. Yanofsky. For a history of DNAX, see Arthur Kornberg, *The Double Helix: Inside Biotech Ventures* (Sausalito, CA: University Science Books, 1995). For an interesting opinion about Dr. Kornberg's book, see Dr. Robert S. Schwartz's review of *The Double Helix: Inside Biotech Ventures* in *NEJM* 333 (1995): 1292–1293.

44. Quotes by Dr. Roncarolo are from an April 2014 Skype interview. Some of these cytokines were independently discovered by other investigators.

45. Information about Dr. Bazan's education and career is from his public LinkedIn profile. https://www.linkedin.com/pub/j-fernando-bazan/4/1a4/948. Also see http://archive.sciencewatch.com/dr/fmf/2010/10julfmf/10julfmfBaza/.

46. Jochen Schmitz et al., "IL-33, an Interleukin-1-like Cytokine that Signals via the IL-1 Receptor-Related Protein ST2 and Induces T Helper Type 2-Associated Cytokines," *Immunity* 23 (2005): 479–490.

47. Ari B. Molofsky, Adam K. Savage, and Richard M. Locksley, "Interleukin-33 in Tissue Homeostasis, Injury, and Inflammation," *Immunity* 42 (2015): 1005–1019; Inna S. Afonina et al., "Proteolytic Processing of Interleukin-1 Family Cytokines: Variations on a Common Theme," *Immunity* 42 (2015): 991–1004.

48. Because maturation of IL-1α is less well understood, and likely more complex, than that of IL1β, I will not discuss it here. For details the reader is referred to Afonina et al., "Proteolytic."

49. Some types of RNA called ribozymes also have enzymatic activity.

50. Douglas Pat Cerretti et al., "Molecular Cloning of the Interleukin-1β Converting Enzyme," *Science* 256 (1992): 97–100; Nancy A. Thornberry et al., "A Novel Heterodimeric Cystein Protease Is Required for Interleukin-1β Processing in Monocytes," *Nature* 356 (1992): 768–774.

51. Production of active IL-1β can at times be independent of Interleukin-1β Converting Enzyme processing. See Giamila Fantuzzi et al., "Response to Local Inflammation of IL-1beta Converting Enzyme-Deficient Mice," *J Immunol* 158 (1997): 1818–1824. Also see Afonina "Proteolytic." None of IL-1β Converting Enzyme's small chemical inhibitors ended up being developed into a drug.

52. For references to the original articles, see Afonina et al., "Proteolytic."

53. Scientists' understanding of cell death, including apoptosis, is constantly evolving, as I will discuss in Chapter 4.

54. We now know of thirteen variants of caspases in humans and three more in other species. Rather ironically, the family member that had started the whole cascade of discoveries—interleukin-1β-converting enzyme—is the one least connected with cell death (it's involved, but in a different way than the other members).

55. For nomenclature, see Emad S. Alnemri et al., "Human ICE/CED-3 Protease Nomenclature," *Cell* 87 (1996): 171. For more information on caspases, see Sonia Shalini et al., "Old, New and Emerging Functions of Caspases," *Cell Death Differ* 22 (2015): 526–539. For more information on apoptosis, see Douglas R. Green, *Means to an End: Apoptosis and Other Cell Death Mechanisms* (Cold Spring Harbor, NY: Cold Spring Harbor Laboratory Press, 2010)

56. Most studies on this issue were—and continue to be—performed on IL-1β, so less is known about IL-18.

57. For references to the original reports, see Taylor A. Doherty, Susannah D. Brydges, and Hal M. Hoffman, "Autoinflammation: Translating Mechanism to Therapy," *J Leukoc Biol* 90 (2011): 37–47.

58. Ibid.

59. Pyroptosis, a type of cell death related to apoptosis, can also be an outcome of inflammasome activation. For more information on inflammasomes, see Haitao Guo, Justin B. Callaway, and Jenny P-Y Ting, "Inflammasomes: Mechanism of Action, Role in Disease, and Therapeutics," *Nat Med* 21 (2015): 677–687.

60. Not all autoinflammatory diseases are caused by mutations in inflammasome components. For more details, see Nienke M. ter Haar et al., "Recommendations for the Management of Autoinflammatory Diseases," *Ann Rheum Dis* 74 (2015): 1636–1644. Newer IL-1 blocking drugs can also be used in autoinflammatory syndromes. In November 2015, the Swiss biochemical company AB2 Bio Ltd. announced use of an IL-18 antagonist called IL-18 binding protein to successfully treat a critically ill baby girl suffering from a severe form of systemic autoinflammation. More information on IL-18 binding protein is presented in Chapter 5. See http://www.ab2bio.com/download/PressRelease_AB2_Bio_November2015_FINAL_EN.pdf.

61. Dr. Bartfai's quotes are from a February 2014 e-mail interview.

62. "The difference between work in industry and academia was small when I started to work," Dr. Bartfai told me. "Back then, the pharmaceutical industry had good research with superb resources, many academic scientists envied those in industry. Today, none of the major drug companies thinks they have the time and funds for any basic research. So, nowadays the difference is huge. If you want to work on basic questions, you have to be in academia. The difference in salaries has also diminished. The employment length at each place is in favor of those in academia, and while they complain about the grant situation in general, an academic scientist who also has teaching has more existential security. But if you love to make new medicines you have to go to industry, as they are the only ones with the know-how and resources."

63. The precise mechanisms of temperature resetting remain somewhat unclear.

64. Elisha Atkins and Pyllis Bodel, "Fever," *NEJM* 286 (1972): 27–34. Prostaglandins

are also essential to maintain the integrity of the mucosa in the gastrointestinal system. Their suppression is the cause of the gut troubles that stem from chronic use of anti-inflammatory drugs.

65. Charles A. Dinarello et al., "Tumor Necrosis Factor (Cachectin) Is an Endogenous Pyrogen and Induces Production of Interleukin 1," *J Exp Med* 163 (1986): 1433–1450; Bruno Conti et al., "Cytokines and Fever," *Front Biosci* 1 (2004): 1433–1449.

CHAPTER 4   Achieving Balance: Pro- and Anti-Inflammatory Mediators

1. For more on this theme, see Sherwin Nuland, "The Uncertain Art: Scatological Medicine," *Am Scholar* 71 (2002): 131–134.

2. Conrad Keating, "Ken Warren and the Rockefeller Foundation's Great Neglected Diseases Network, 1978–1988: The Transformation of Tropical and Global Medicine," *Mol Med* 20 (2014): S24–S30. For a rather different perspective on the Rockefeller Foundation, see E. Richard Brown, *Rockefeller Medicine Men: Medicine and Capitalism in the Progressive Era* (Berkeley: University of California Press, 1979).

3. Anthony Cerami, "The Value of Failure: the Discovery of TNF and Its Natural Inhibitor Erythropoietin," *J Internal Med* 269 (2011): 8–15.

4. However, for reasons unclear to me, the researchers chose not to use the term *cachectin* in their publications until 1984, preferring instead the more generic "endotoxin-induced mediator."

5. Cerami, "The Value." This is the same Bruce Beutler who received the Nobel Prize in 2011, as mentioned in Chapter 2.

6. Siddhartha Muhkerjee, *The Emperor of All Maladies: A Biography of Cancer* (New York: Scribner, 2010), 466 (Kindle edition).

7. William B. Coley, "The Treatment of Malignant Tumors by Repeated Inoculations of Erysipelas. With a Report of Ten Original Cases," *Am J Med Sci* 105 (1893): 487–494.

8. For an example of this debate, see John J. L. Playfair, Janice Taverne, and Nick Matthews, "What Is Tumor Necrosis Factor Really For?" *Immunol Today* 5(1984): 165–166. In some of their studies, Dr. Old and other researchers used the same experimental approach that was later instrumental in discovering IL-18, that is, administration of *P. acnes* (at the time known as *C. parvum*) followed by an injection of endotoxin (see Chapter 3).

9. The quote is from Bharat B. Aggarwal, Subash C. Gupta, and Ji Hye Kim, "Historical Perspectives on Tumor Necrosis Factor and Its Superfamily: 25 Years Later, a Golden Journey," *Blood* 119 (2012): 651–665. For more on the discovery of lymphokines and their role in cancer therapy, see Stephen S. Hall, *A Commotion in the Blood: Life, Death, and the Immune System* (New York: Henry Holt, 1997). For more on Genentech, see Sally Smith Hughes, *Genentech: The Beginnings of Biotech* (Chicago: University of Chicago Press, 2011).

10. The early cytokine nomenclature included terms such as lymphokines for cytokines supposedly produced only by lymphocytes and monokines for those that would have monocytes as their only source. This terminology was

later for the most part abandoned when it became clear that most cytokines can be produced by different types of cells.

11. References to the original articles can be found in Aggarwal, Gupta, and Kim, "Historical." Also see Walter Fiers et al., "Gene Cloning and Structure-Function Relationship of Cytokines Such as TNF and Interleukins," *Immunol Lett* 16 (1987): 219–226; Lloyd J. Old, "Tumor Necrosis Factor (TNF)," *Science* 239 (1985): 630–632.

12. For references to the original publications, see Cerami, "The Value."

13. For a first-person account of these and other studies on TNF in sepsis and shock, as well as their clinical implications, see Kevin J. Tracey, *Fatal Sequence: The Killer Within* (New York: Dana Press, 2005). Dr. Tracey was the first author of the publication using baboons; he continues to combine his clinical work with research in inflammation.

14. Andrew R. Exley et al., "Monoclonal Antibody to TNF in Severe Septic Shock," *The Lancet* 335 (1990): 1275–1277.

15. Edward Abraham et al., "Efficacy and Safety of Monoclonal Antibody to Human Tumor Necrosis Factor Alpha in Patients with Sepsis Syndrome. A Randomized, Controlled, Double-Blind, Multicenter Clinical Trial. TNF-Alpha Mab Sepsis Study Group," *JAMA* 273 (1995): 934–941; Steven M. Opal et al., "Confirmatory Interleukin-1 Receptor Antagonist Trial in Severe Sepsis: a Phase III, Randomized, Double-Blind, Placebo-Controlled, Multicenter Trial. The Interleukin-1 Receptor Antagonist Sepsis Investigator Group," *Crit Care Med* 25 (1997): 1115–1124. Other clinical trials on anti-TNF had been initiated and later either confirmed lack of effects or were stopped early.

16. See, for example, Linda Kirschenbaum, Mark Astiz, and Eric C. Rackow, "Antibodies to TNF-alpha: Too Little, Too Late?" *Crit Care Med* 26 (1998): 1625–1626; Georges E. Grau and Daniela N. Maennel, "TNF Inhibition and Sepsis—Sounding a Cautionary Note," *Nat Med* 3 (1997): 1192–1195.

17. Mark K. Eskandari et al., "Anti-Tumor Necrosis Factor Antibody Therapy Fails to Prevent Lethality after Cecal Ligation and Puncture or Endotoxemia," *J Immunol* 148 (1992): 2724–2730. These authors also failed to demonstrate protection by anti-TNF treatment when using the endotoxic shock model. For more on the ongoing debate about animal models, see the long list of articles at http://www.ncbi.nlm.nih.gov/pubmed/?term=23401516.

18. Kirschenbaum et al., "Antibodies." Also see, Jan Vilcek, "First Demonstration of the Role of TNF in the Pathogenesis of Disease," *J Immunol* 181 (2006): 5–6.

19. Edward Abraham, "Why Immunomodulatory Therapies Have Not Worked in Sepsis," *Inten Care Med* 25 (1999): 556–566.

20. Julius Cohnheim, *Lectures on General Pathology. A Handbook for Practitioners and Students. Section II. The Pathology of Nutrition*, trans. from the German 2nd ed. by Alexander B. McKee (London: The New Sydenham Society, 1889–90), 549, 551, 572–573.

21. A third type of cell death is called autophagic.

22. Lorenzo Galluzzi et al., "Cell Death Modalities: Classification and Pathophysiological Implications," *Cell Death Differ* 14 (2007): 1237–1266.

23. Lorenzo Galluzzi et al., "Essential versus Accessory Aspects of Cell Death: Recommendations of the NCCD 2015," *Cell Death Differ* 22 (2015): 58–73.

24. Elizabeth A. Carswell et al., "An Endotoxin-Induced Serum Factor That Causes Necrosis of Tumors," *PNAS* 72 (1975): 3666–3670.

25. Bharat B. Aggarwal et al., "Human Tumor Necrosis Factor. Production, Purification, and Characterization," *J Biol Chem* 26 (1985): 2345–2354. For more on the history of necrosis and cell death, see David Wallach, Tae-Bang Kang, and Andrew Kovalenko, "Concepts of Tissue Injury and Cell Death in Inflammation: A Historical Perspective," *Nat Rev Immunol* 14 (2014): 51–59. Fibroblasts are cells that contribute to form the structure of tissues and are one of the main cellular components of connective tissue, which supports and connects other types of tissues.

26. Skype interview, January 2014.

27. For an interesting article on concepts related to the study and interpretation of cell death, see Andrew S. Reynolds, "The Deaths of a Cell: How Language and Metaphor Influence the Science of Cell Death," *Stud Hist Philos Biol Biomed Sci* 28 (2014): 175–184.

28. The quote is from a January 2014, Skype interview. See also David Wallach, "The TNF Cytokine Family: One Track in a Road Paved by Many," *Cytokine* 63 (2013): 225–229.

29. Jiahuai Han, Chuan-Qi Zhong, and Duan-Wu Zhang, "Programmed Necrosis: Backup to and Competitor with Apoptosis in the Immune System," *Nat Immunol* 12 (2011): 1143–1149; Deborah A. Flusberg and Peter K. Sorger, "Surviving Apoptosis: Life-Death Signaling in Single Cells," *Trends Cell Biol* 25 (2015): 446–458; Daniel J. Van Antwerp et al., "Suppression of TNF-α-Induced Apoptosis by NF-κB," *Science* 274 (1996): 787–789.

30. For more information on the use of TNF in melanoma, see Jan Deroose et al., "Isolated Limb Perfusion for Melanoma In-Transit Metastases: Development in Recent Years and the Role of Tumor Necrosis Factor Alpha," *Curr Opin Oncol* 23 (2011): 183–188. For more on the role of TNF in cancer, see Toru Atsumi et al., "Inflammation Amplifier, A New Paradigm in Cancer Biology," *Cancer Res* 74 (2014): 8–14.

31. Wenjun Ouyang et al., "Regulation and Functions of the IL-10 Family of Cytokines in Inflammation and Disease," *Annu Rev Immunol* 29 (2011): 71–109.

32. Tim R. Mosmann and Robert L. Coffman, "Two Types of Mouse Helper T-Cell Clone. Implications for Immune Regulation," *Immunol Today* 8 (1987): 223–227.

33. What if you are infected with both mycobacteria (or trypanosomes) and worms? After all, this is a very common situation in many parts of the world, and has likely been so in most places for much of human evolutionary history. Here things can get complicated. We know, for example, that a worm infestation weakens the ability of the body to fight against tuberculosis and may reduce effectiveness of antituberculosis interventions. For more details see Xin-Xu Li and Xiao-Nong Zhou, "Co-Infection of Tuberculosis and Parasitic Diseases in Humans: A Systematic Review," *Parasit Vectors* 6 (2013): 79. Scientists now recognize more than two types of T helper cells and

appreciate the importance of T helper lymphocytes in many other types of diseases. For more information, see Francesco Annunziato, Chiara Romagnani, and Sergio Romagnani, "The 3 Major Types of Innate and Adaptive Cell-Mediated Effector Immunity," *J Allergy Clin Immun* 135 (2015): 626–635.

34. For references to the original articles, see Kevin W. Moore et al., "Interleukin-10," *Annu Rev Immunol* 11 (1993): 165–190.

35. For references to the original articles, see ibid. After nine years at DNAX, Dr. Roncarolo moved back to Italy, where she directed the Telethon Institute for Gene Therapy, near Milano, and was then director of research of the San Raffaele Scientific Institute, one of Italy's premier research centers. She has recently returned to California, where she codirects the Institute for Stem Cell Biology and Regenerative Medicine at Stanford University.

36. It is interesting to consider this change in terminology within the framework of societal changes from a manufactory-based to an information-centered environment discussed by Hannah Landecker in her article "Postindustrial Metabolism: Fat Knowledge," *Public Culture* 25 (2013): 495–522.

37. Fiona M. Powrie and Don Mason, "OX-22^{high} CD4$^+$ T Cells Induce Wasting Disease with Multiple Organ Pathology: Prevention by the OX-22^{low} Subset," *J Exp Med* 172 (1990): 1701–1708.

38. Amy Maxmen, "Fiona Powrie: Gut Diplomacy," *J Exp Med* 207 (2010): 4–5.

39. Hervé Groux et al., "A CD4$^+$ T-Cell Subset Inhibits Antigen-Specific T-Cell Responses and Prevents Colitis," *Nature* 389 (1997): 737–742; Maria Grazia Roncarolo et al., "Tr1 Cells and the Counter-Regulation of Immunity: Natural Mechanisms and Therapeutic Applications," *Curr Top Microbiol Immunol* 380 (2014): 39–68; Julia Bollrath and Fiona M. Powrie, "Controlling the Frontier: Regulatory T-Cells and Intestinal Homeostasis," *Semin Immunol* 25 (2013): 352–357.

40. For example, see Xuan Gao et al., "Interleukin-10 Promoter Gene Polymorphisms and Susceptibility to Tuberculosis: A Meta-Analysis," *PLoSOne* 10 (2015): e0127496 and references therein. Despite the path that led to its identification, IL-10 turned out not be the only Th2 cytokine that inhibits production of interferon-γ, and probably not even the most important one. For more information, see Stefanie Eyerich and Christina E. Zielinski, "Defining Th-cell Subsets in a Classical and Tissue-Specific Manner: Examples from the Skin," *Eur J Immunol* 44 (2014): 3475-3483.

41. For example, see Sarah C. Higgins et al., "Toll-Like Receptor-4-Mediated Innate IL-10 Activates Antigen-Specific Regulatory T Cells and Confers Resistance to *Bordetella Pertussis* by Inhibiting Inflammatory Pathology," *J Immunol* 171 (2003): 3119–3127, and references therein.

42. Christopher A. Hunter et al., "IL-10 Is Required to Prevent Immune Hyperactivity During Infection with *Trypanosoma cruzi*," *J Immunol* 158 (1997): 3311–3316. Also see Ouyang, "Regulation."

43. Julius Cohnheim, *Lectures on General Pathology. A Handbook for Practitioners and Students. Section I. The Pathology of the Circulation*, trans. from the German 2nd ed. by Alexander B. McKee (London: The New Sydenham Society, 1889), 250.

44. Quotes are from a January 2014 Skype interview and from https://ncifrederick .cancer.gov/events/oppenheim/wellWishes.asp.

45. For references to the original articles, see Kouji Matsushima and Joost J. Oppenheim, "Interleukin 8 and MCAF: Novel Inflammatory Cytokines Inducible by IL-1 and TNF," *Cytokine* 1 (1989): 2–13. For accounts of the discovery of IL-8 and MCP-1 by some of the other researchers involved, see Teizo Yoshimura, "Discovery of IL-8/CXCL8 (The Story from Frederick)," *Front Immunol* 6 (2015): 00278; Marco Baggiolini, "CXCL8—The First Chemokine," *Front Immunol* 6 (2015): 00285. About the reasons for classifying IL-8 as an interleukin, during our interview Dr. Oppenheim explained, "We had a little meeting, the investigators who had cloned it, and Ed Leonard favored the name NAP—neutrophil-activating peptide—for the factor, but at that time I thought the molecule also acted on lymphocytes to some extent, so I said we should call it IL-8 because that would call much more attention to it, as a way of advertising this."

46. See Albert Zlotnick and Osamu Yoshie, "The Chemokine Superfamily Revisited," *Immunity* 36 (2012): 705–716.

47. Ibid.

48. For more on the connection between chemokine receptors and microbes, see Richard Horuk, "The Duffy Antigen Receptor for Chemokines DARC/ ACKR1," *Front Immunol* 6 (2015): 00279; Philip A. Murphy, "Viral Chemokine Receptors," *Front Immunol* 6 (2015): 00281; Paolo Lusso, "Chemokines and HIV: The First Close Encounter," *Front Immunol* 6 (2015): 00294; Marc Parmentier, "CCR5 and HIV Infection, a View from Brussels," *Front Immunol* 6 (2015): 00295.

49. Silvano Sozzani et al., "Chemokines as Effector and Target Molecules in Vascular Biology," *Cardiovasc Res* 107 (2015): 364–372.

50. Quotes from Dr. Flower are from a February 2014 e-mail interview.

51. Dr. Vane's Nobel lecture can be accessed at http://www.nobelprize.org /nobel_prizes/medicine/laureates/1982/vane-lecture.html. As acknowledged in the lecture, Drs. Smith and Willis had simultaneously arrived at the same conclusion and their article was published in the same issue of *Nature* as Dr. Vane's reports.

52. Dr. Otto Loewi was a German physician and scientist who shared with Dr. Henry H. Dale the 1936 Nobel Prize in Physiology or Medicine for discovering chemical mediators involved in transmission of nerve impulses, chiefly the neurotransmitter acetylcholine.

53. Roderick R. Flower and George J Blackwell, "Anti-Inflammatory Steroids Induce Biosynthesis of a Phospholipase A2 Inhibitor Which Prevents Prostaglandin Generation, *Nature* 278 (1979): 456–459.

54. Massimo Di Rosa et al., "Nomenclature Announcement. Anti-Phospholipase Proteins," *Prostaglandins* 28 (1984): 441–442.

55. Barbara P. Wallner et al., "Cloning and Expression of Human Lipocortin, a Phospholipase A2 Inhibitor with Potential Anti-Inflammatory Activity," *Nature* 320 (1986): 77–81; Roderick J. Flower, "The Mediators of Steroid Action," *Nature* 320 (1986): 20.

56. Bruce Beutler and Roderick J. Flower, "Consensus and Confusion Among the Contradas," *New Biologist* 3 (1990): 249–253.

57. Also see Roderick J. Flower, "On the Eicosanoid Trail with John Vane and Jack McGiff: 1974–1976. A Personal Reminiscence," *Prostag Oth Lipid M*, 120 (2015): 3-8.

58. After complex restructuring, Sclavo exclusively returned to its core mission of vaccine development. Directed by Dr. Rino Rappuoli, the institute in Siena remains one of the world's premier centers for research on vaccines. Much research on IL-1 and other cytokines was performed at Sclavo, with Drs. Diana Boraschi and Aldo Tagliabue playing prominent roles.

59. Dr. Perretti's wife is Dr. Marzia Malcangio, a professor at the Wolfson Centre for Age-Related Diseases of King's College London.

60. Mauro Perretti and Roderick J. Flower, "Modulation of IL-1-Induced Neutrophil Migration by Dexamethasone and Lipocortin 1," *J Immunol* 150 (1993): 992–999.

61. Mauro Perretti et al., "Mobilizing Lipocortin 1 in Adherent Human Leukocytes Downregulates Their Transmigration," *Nat Med* 2 (1996): 1259–1262. Recall that the term *autocrine* refers to the action of a body message that feeds back on the cell that had produced and released it. In 1996, Dr. Perretti obtained a PhD in London after deciding not to return to Italy.

62. Antje Walther, Kristina Riehemann, and Volker Gerke, "A Novel Ligand of the Formyl Peptide Receptor: Annexin I Regulates Neutrophil Extravasation by Interacting with the FPR," *Mol Cell* 5 (2000): 831–840; Mauro Perretti et al., "Involvement of the Receptor of Formylated Peptides in the *In Vivo* Anti-Migratory Actions of Annexin 1 and Its Mimetics," *Am J Pathol* 158 (2001): 1969–1973.

63. For more information on the resolution of inflammation, see Mauro Perretti, ed., *Resolution of Inflammation. Semin Immunol* 27 (2015): 145–234.

64. Mauro Perretti et al., "Endogenous Lipid- and Peptide-Derived Anti-Inflammatory Pathways Generated with Glucocorticoids and Aspirin Treatment Activate the Lipoxin A4 Receptor," *Nat Med* 8 (2002): 1296–1302. The same formyl peptide receptor binds additional molecules, some of which exert pro-inflammatory effects, making the precise mechanisms through which this receptor participates in regulation of inflammation quite arduous to understand.

65. Thomas T. MacDonald et al., "Tumour Necrosis Factor-Alpha and Interferon-Gamma Production Measured at the Single Cell Level in Normal and Inflamed Human Intestine," *Clin Exp Immunol* 81 (1990): 301–305; Christian P. Braegger et al., "Tumour Necrosis Factor Alpha in Stool as a Marker of Intestinal Inflammation," *The Lancet* 339 (1992): 89–91; Hendrik M. van Dullemen et al., "Treatment of Crohn's Disease With Anti-Tumor Necrosis Factor Chimeric Monoclonal Antibody (cA2)," *Gastroenterology* 109 (1995): 129–135. For details about the development of Infliximab, see Jan Vilcek, *Love and Science: A Memoir* (New York; Seven Stories Press, 2016).

66. For references to original articles, see Simon T. C. Peake et al., "Mechanisms of Action of Anti-Tumor Necrosis Factor α Agents in Crohn's Disease," *Inflamm Bowel Dis* 19 (2013): 1546–1555. There is evidence that anti-TNF therapy can not only treat IBD but also have the exact opposite effect, that is,

elicit or precipitate this condition when used as a medication for other diseases. See, for example, Boulos Haraoui, and Marilyn Krelenbaum, "Emergence of Crohn's Disease During Treatment With the Anti-Tumor Necrosis Factor Agent Etanercept for Ankylosing Spondylitis: Possible Mechanisms of Action," *Sem Arthritis Rheum* 39 (2009): 176-181, and references therein.

67. Moritz Leppkes et al., "Pleiotropic Functions of TNF-α in the Regulation of the Intestinal Epithelial Response to Inflammation," *Int Immunol* 26 (2014): 598–515; David A. Leiman and Gary R. Lichtenstein, "Therapy of Inflammatory Bowel Disease: What to Expect in the Next Decade," *Curr Opin Gastroen* 30 (2014): 385–390; D. M. Knight et al., "Construction and Initial Characterization of a Mouse-Human Chimeric Anti-TNF Antibody," *Mol Immunol* 30 (1993): 1443–1453. Though I interviewed Dr. Vilcek for this book, readers will definitely be better off with his first-person account of his interesting life in *Love and Science: A Memoir*.

68. For references to the original articles, see Dror S. Shouval et al., "Interleukin 10 Receptor Signaling: Master Regulator of Intestinal Mucosal Homeostasis in Mice and Humans," *Adv Immunol* 122 (2014): 177–210.

69. Gareth J. Marlow, Dominique van Gent, and Lynnette R. Ferguson, "Why Interleukin-10 Supplementation Does Not Work in Crohn's Disease Patients," *World J Gastroenterol* 19 (2013): 3931–3941.

70. Youichirou Sakanoue et al., "Lipocortin-Present Perforating and Lipocortin-Absent Nonperforating Crohn's Disease, *Am J Surg* 164 (1992): 341–344; Giovanna Leoni et al., "Annexin A1-Containing Extracellular Vesicles and Polymeric Nanoparticles Promote Epithelial Wound Repair," *J Clin Invest* 125 (2015): 1215–1227; Nengtai Ouyang et al., "MC-12, an Annexin A1-Based Peptide, Is Effective in the Treatment of Experimental Colitis," *PLOS One* 7 (2012): e41585.

CHAPTER 5   Traps and Diversions: Cytokine Decoys

1. These devices still exist, in modified form, and are still called decoys.

2. Ralph Payne-Gallway, *The Book of Duck Decoys: Their Construction, Management, and History* (London: J. Van Voorst, 1886), 3.

3. Samuel Bogoch, Robert F. Gilfillan, and Paul Evans, "Sialoresponsin: Neuraminic Acid-Containing Substance(s) Which Accumulates in Chorioallantoic Fluid During the First Minutes of Virus Infection," *Nature* 196 (1962): 649–651.

4. Payne-Gallway, *Duck Decoys*, 3.

5. For an analysis of the reasons for the early rejection of chromatography, see Jonathan Livengood, "Why Was M. S. Tswett's Chromatographic Adsorption Analysis Rejected?" *Stud Hist Phily Sci Part A* 40 (2009): 57–69. For more on the history of chromatography, see *Chromatography: A Century of Discovery 1900–2000*, eds. Charles W. Gehrke, Robert L. Wixom, and Ernst Bayer (Amsterdam, The Netherlands: Elsevier Science, 2001).

6. The quote is from http://www.explainthatstuff.com/chromatography.html.

7. Philippe Seckinger et al., "A Urine Inhibitor of Interleukin 1 Activity the Blocks Ligand Binding," *J Immunol* 139 (1987): 1546–1549.

8. For technical details on affinity chromatography specifically in the context of cytokine research, see Daniela Novick and Menachem Rubinstein, "Ligand Affinity Chromatography, an Indispensable Method for the Purification of Soluble Cytokine Receptors and Binding Proteins," *Method Mol Biol* 820 (2012): 195–214.

9. Dr. Novick's quotes and personal information are from a January 2014 e-mail interview. Rebif™ is a recombinant version of Interferon-β1a used for treatment of the autoimmune disease multiple sclerosis. Enbrel™ is a soluble TNF receptor used by patients with rheumatoid arthritis, psoriasis, and other conditions. The following pages will provide more details on this drug. During the interview, Dr. Novick also stated, "As a woman in a field mostly dominated by men, I encountered many problems. Women do not count unless they are assertive yet have charm, unless everybody around knows they mean business, unless they work extremely hard both for their career and raising a family, unless they are successful all along. When I felt deep inside to be right, whether in my research or referring to my promotion, I went through the wall to achieve it, disregarding what was being said around and sometimes risking my position. In retrospect, it always proved itself. I published over one hundred research papers, wrote reviews and book chapters, and I hold patents. In 2005, I was a corecipient of the Milstein award in recognition of exceptional contributions to research related to interferons and cytokines."

10. For references to the original articles and a history of these discoveries, see Daniela Novick and Menachem Rubinstein, "The Tale of Soluble Receptors and Binding Proteins: From Bench to Bedside," *Cytokine Growth F R* 18 (2007): 525–533. See also Daniela Novick et al., "The Human Interferon-γ Receptor. Purification, Characterization, and Preparation of Antibodies," *J Biol Chem* 262 (1987): 8483–8487; Michel Aguet and Gilles Merlin, "Purification of Human γ Interferon Receptors by Sequential Affinity Chromatography on Immobilized Monoclonal Antireceptor Antibodies and Human γ Interferon," *J Exp Med* 165 (1987): 988–999.

11. This quote is from a Skype conversation with Dr. Wallach, August 3, 2015.

12. Philippe Seckinger, Sylvia Isaaz, and Jean-Michel Dayer, "A Human Inhibitor of Tumor Necrosis Factor α," *J Exp Med* 167 (1988): 1511–1516.

13. Although I am using the term decoy here, this terminology was not introduced until 1993 and first used in reference to a different molecule.

14. Christina Peetre et al., "A Tumor Necrosis Factor Binding Protein Is Present in Human Biological Fluids," *Eur J Haematol* 41 (1988): 414–419; Inge Olsson et al., "Isolation and Characterization of a Tumor Necrosis Factor Binding Protein from Urine," *Eur J Haematol* 42 (1989): 270–275.

15. The quote is from a January 2014 interview with Dr. Wallach.

16. Hartmut Engelmann et al., "A Tumor Necrosis Factor-binding Protein Purified to Homogeneity from Human Urine Protects Cells from Tumor Necrosis Factor Toxicity," *J Biol Chem* 264 (1989): 11974–11980. Although Dr. Novick is not a coauthor in this article, she is acknowledged for her advice. See also, David Wallach, "The TNF Cytokine Family: One Track in a Road Paved by Many," *Cytokine* 63 (2013): 225–229.

17. Yaron Nophar et al., "Soluble Forms of Tumor Necrosis Factor Receptors (TNF-Rs). The cDNA for the Type I TNF-R, Cloned Using Amino Acid Sequence Data of its Soluble Form, Encodes Both the Cell Surface and a Soluble Form of the Receptor," *EMBO J* 9 (1990): 3269–3278.

18. The original neutrophil shedding article is Françoise Porteu and Carl Nathan, "Shedding of Tumor Necrosis Factor Receptors by Activated Human Neutrophils," *J Exp Med* 172 (1990): 599–607. For more information on receptor shedding, see Rama Khokha, Aditya Murthy, and Ashley Weiss, "Metalloproteinases and Their Natural Inhibitors in Inflammation and Immunity," *Nat Rev Immunol* 13 (2013): 649–665. Actually, TNF itself is produced as a membrane-bound protein that needs to be clipped before it can turn into a soluble cytokine. The very first clipping enzyme discovered was the one that cuts TNF away from the cell surface. Initially called TACE, the enzyme was later renamed ADAM17 and shown to clip many surface proteins, including TNF's own receptors. However, not every soluble receptor is generated by shedding, some being produced already as soluble molecules by a process called alternative splicing.

19. The quote is from Novick and Rubinstein, "The Tale." Hartmut Engelmann, Daniela Novick, and David Wallach, "Two Tumor Necrosis Factor-Binding Proteins Purified from Human Urine," *J Biol Chem* 265 (1990): 1531–1536.

20. Tadahiko Kohno et al., "A Second Tumor Necrosis Factor Receptor Gene Product Can Shed a Naturally Occurring Tumor Necrosis Factor Inhibitor," *PNAS* 87 (1990): 8331–8335.

21. Charles A. Dinarello et al., "Tumor Necrosis Factor (Cachectin) Is an Endogenous Pyrogen and Induces Production of Interleukin 1," *J Exp Med* 163 (1986): 1433–1450; Philip Stashenko et al., "Synergistic Interactions Between Interleukin 1, Tumor Necrosis Factor, and Lymphotoxin in Bone Resorption," *J Immunol* 138 (1987): 1464–1468.

22. Jean-Michel Dayer, Bruce Beutler, and Anthony Cerami, "Cachectin/Tumor Necrosis Factor Stimulates Collagenase and Prostaglandin E2 Production by Human Synovial Cells and Dermal Fibroblasts," *J Exp Med* 162 (1985): 2163–2168; Jeremy Saklatvala, "Tumour Necrosis Factor Alpha Stimulates Resorption and Inhibits Synthesis of Proteoglycan in Cartilage," *Nature* 322 (1986): 547–549; Fionula M. Brennan et al., "Inhibitory Effect of TNF alpha Antibodies on Synovial Cell Interleukin-1 Production in Rheumatoid Arthritis," *Lancet* 29 (1989): 244–247.

23. Michael J. Elliott et al., "Treatment of Rheumatoid Arthritis with Chimeric Monoclonal Antibodies to Tumor Necrosis Factor Alpha," *Arthritis Rheum* 36 (1993): 1681–1690; Michael J. Elliott et al., "Randomised Double-Blind Comparison of Chimeric Monoclonal Antibody to Tumor Necrosis Factor Alpha (cA2) versus Placebo in Rheumatoid Arthritis," *Lancet* 344 (1994): 1105:1110. It wasn't until 2001 that the Food and Drug Administration approved IL-1Ra (Anakinra) for treatment of rheumatoid arthritis. See also, http://www.laskerfoundation.org/awards/2003_c_description.htm.

24. Karsten Peppel, David Crawford, and Bruce Beutler, "A Tumor Necrosis Factor (TNF) Receptor-IgG Heavy Chain Chimeric Protein as a Bivalent Antagonist of TNF Activity," *J Exp Med* 174 (1991): 1483–1489. Because

biology is never as neat as we abstractly imagine it to be, both forms of TBP actually serve a dual, contrasting role. Whereas at high concentrations TBPs inhibit TNF's activity, acting as true decoys, the work of Dr. Dan Aderka in Dr. Wallach's group demonstrated that low levels of TBPs stabilize TNF, thus prolonging and increasing—not decreasing—TNF's activity. Although the evolutionary meaning of this observation remains unclear, this anti-decoy effect of TBP is definitely not something desirable in a drug aimed at blocking TNF. Bruce Beutler's molecular chimera would thus also overcome this potentially troublesome aspect of TBP's biology. See Dan Aderka et al., "Stabilization of the Bioactivity of Tumor Necrosis Factor by Its Soluble Receptors," *J Exp Med* 175 (1992): 323–329. In a conversation (August 3, 2015), Dr. Wallach told me that accumulating evidence has made him believe that endogenous TNF soluble receptors may contribute to disease, even when present at high levels, by stabilizing TNF and preventing its clearance.

25. Larry W. Moreland et al., "Etanercept Therapy in Rheumatoid Arthritis. A Randomized, Controlled Trial," *Ann Intern Med* 130 (1999): 478–486. Investigators tested the efficacy of Etanercept in septic shock, just as they had done with anti-TNF antibodies and IL1Ra, but this type of treatment also failed. Charles J. Fisher et al., "Treatment of Septic Shock with the Tumor Necrosis Factor Receptor:Fc Fusion Protein. The Soluble Receptor Sepsis Study Group," *NEJM* 334 (1996): 1697–1702. Immunex also made a chimeric version of soluble TNF receptor type I, the original version of TBP; it is called Onercept and was evaluated in clinical trials but not brought to the market. However, other drugs made through the same recombinant technology were later approved, including Afibercept for treatment of a type of macular degeneration, a disease of the eye.

26. William J. Sandborn et al., "Etanercept for Active Crohn's Disease: A Randomized, Double-Blind, Placebo-Controlled Trial," *Gastroenterology* 121 (2001): 1088–1094.

27. For references to the original articles, see Simon T. C. Peake et al., "Mechanisms of Action of Anti-Tumor Necrosis Factor α Agents in Crohn's Disease," *Inflamm Bowel Dis* 19 (2013): 1546–1555; Paolo Biancheri et al., "Proteolytic Cleavage and Loss of Function of Biologic Agents That Neutralize Tumor Necrosis Factor in the Mucosa of Patients with Inflammatory Bowel Disease," *Gastroenterology* 149 (2015): 1564–1574.

28. Alex Philippidis, "The Top 25 Best-Selling Drugs of 2014," *Engineering and Biotechnology News*, February 23, 2015. There are wide ranges of estimates for the cost of treatment with anti-TNF drugs that depend on the type of analysis performed as well as the country involved.

29. Because the ovaries' function declines after menopause, the pituitary gland raises gonadotropin levels in an intensifying attempt to release the remaining eggs.

30. Bruno Lunenfeld, "Management of Infertility: Past, Present and Future (From a Personal Perspective)," *Journal für Reproduktionsmedizin und Endokrinologie* 10 (2013): 13–22. For more on the history of Serono, see http://www.encyclopedia.com/doc/1G2-2845100098.html. The critical involvement of Dr. Lunenfeld in development of Pergonal led to an agreement with Serono

that provided Israeli women with free access to the drug for several years. See http://www.israel21c.org/people/baby-boomer/.

31. Skype interview with Dr. Wallach, January 2014.

32. Though this is not the place for discussing such issues, the heavy involvement of the Catholic Church in the process of development of drugs that are essential for *in vitro* fertilization—a procedure still officially condemned by the Vatican—deserves consideration, as does the close relationship of a then Vatican-owned company with Israeli institutions.

33. All quotes and personal information from Dr. Mantovani are from a July 2015 Skype interview.

34. In the Italian system one moves directly from high school to specialized university studies, without the intermediate years of undergraduate college.

35. As the first question of the final exam, Dr. Clerici, the immunology professor, produced a urine sample from his sister-in-law and asked Mantovani to perform a pregnancy test. This answer, and the subsequent ones, being correct, Dr. Mantovani passed the exam.

36. Information on the history of the Mario Negri Institute is from http://www.marionegri.it/en_US/home/istituto_en/chi_siamo_en/storia_en. I was at the Mario Negri Institute when Dr. Mantovani was director of Immunology, though I did not study or work directly under his mentorship.

37. Elisabetta Dejana et al., "Stimulation of Prostacyclin Synthesis in Vascular Cells by Mononuclear Cell Products," *Blood* 64 (1984): 1280–1283; Vincenzo Rossi et al., "Prostacyclin Synthesis Induced in Vascular Cells by Interleukin-1," *Science* 229 (1985): 174–176.

38. Francesco Colotta et al., "Expression of c-fos Protooncogene in Normal Human Peripheral Blood Granulocytes," *J Exp Med* 165 (1987): 1224–1229.

39. Francesco Colotta et al., "Modulation of Granulocyte Survival and Programmed Cell Death by Cytokines and Bacterial Products," *Blood* 80 (1992): 2012–2020.

40. John E. Sims et al., "cDNA Expression Cloning of the IL-1 Receptor, a Member of the Immunoglobulin Superfamily," *Science* 241 (1988): 585–589; Catherine J. McMahan et al., "A Novel IL-1 Receptor, Cloned from B Cells by Mammalian Expression, Is Expressed in Many Cell Types," *EMBO J* 10 (1991): 2821–2832; Eleni Stylianou et al., "Interleukin 1 Induces NF-κB Through Its Type I but Not Its Type II Receptor in Lymphocytes," *J Biol Chem* 267 (1992): 15836–15841.

41. Julie A. Eastgate, Julian A. Symons, and Gordon W. Duff, "Identification of an Interleukin-1 beta Binding Protein in Human Plasma," *FEBS Lett* 260 (1990): 213–216; Julian A. Symons and Gordon W. Duff, "A Soluble Form of the Interleukin-1 Receptor Produced by a Human B Cell Line," *FEBS Lett* 272 (1990): 133–136; Judith G. Giri, Robert C. Newton, and Richard Horuk, "Identification of Soluble Interleukin-1 Binding Protein in Cell-free Supernatants," *J Biol Chem* 265 (1990): 17416–17419; Julian A. Symons, Julie A. Eastgate, and Gordon W. Duff, "Purification and Characterization of a Novel Soluble Receptor for Interleukin 1," *J Exp Med* 174 (1991): 1251–1254.

42. John E. Sims et al., "Interleukin 1 Signaling Occurs Exclusively via the Type I Receptor," *PNAS* 90 (1993): 6155–6159.

43. Francesco Colotta et al., "Interleukin 1 Type II Receptor: A Decoy Target for

IL-1 That Is Regulated by IL-4," *Science* 261 (1993): 472–475; Julian A. Symons, Peter R. Young, and Gordon W. Duff, "Soluble Type II Interleukin 1 (IL-1) Receptor Binds and Blocks Processing of IL-1β and Loses Affinity for IL-1 Receptor Antagonist," *PNAS* 92 (1995): 1714–1718. For reference to subsequent studies, see Vanessa A. Peters, Jennifer J. Joesting, and Gregory G. Freund, "IL-1 Receptor 2 (IL-1R2) and Its Role in Immune Regulation," *Brain Behav Immun* 32 (2013): 1–8.

44. Dr. Novick's quote is from Novick and Rubinstein, "The Tale." Daniela Novick, Batya Cohen, and Menachem Rubinstein, "Soluble Interferon-alpha Receptor Molecules Are Present in Body Fluids," *FEBS Lett* 314 (1992): 445–448; Daniela Novick, Batya Cohen, and Menachem Rubinstein, "The Human Interferon α / β Receptor: Characterization and Molecular Cloning," *Cell* 77 (1994): 391–400; Daniela Novick et al., "Soluble and Membrane-Anchored Forms of the Human IFN-alpha/beta Receptor," *J Leukoc Biol* 57 (1995): 712–718; Batya Cohen et al., "Ligand-Induced Association of the Type I Interferon Receptor Components," *Mol Cell Biol* 15 (1995): 4208–4214. For more information on Type I Interferon, its receptor, and the history of their discovery, see Jan Vilcek, "My Fifty Years with Interferon," *J Interf Cytok Res* 27 (2007): 535–541; Jan Vilcek, "Interferon Research BC (Before Cloning)," *Curr Top Microbiol Immunol* 316 (2007): 9–22; Christopher Krause and Sidney Pestka, "Historical Developments in the Research of Interferon Receptors," *Cytokine Growth F R* 18 (2007): 473–482

45. Lawrence M. Pfeffer et al., "The Short Form of the Interferon α/β Receptor Chain 2 Acts as a Dominant Negative for Type I Interferon Action," *J Biol Chem* 272 (1997): 11002–11005.

46. I was part of Dr. Dinarello's group at the time, each member of the team working on various aspects of IL-18's biology. While Dr. Adrian Puren used *in vitro* techniques to demonstrate the plentiful pro-inflammatory effects of IL-18, I mostly employed *in vivo* animal models to examine the regulation of IL-18 production and effects, other researchers examining additional aspects of this cytokine's biology. See, for example, Adrian Puren et al., "Interleukin-18 (IFNgamma-Inducing Factor) Induces IL-8 and IL-1beta via TNFalpha Production from non-CD14+ Human Blood Mononuclear Cells," *J Clin Invest* 101 (1998): 711–721; Giamila Fantuzzi et al., "Interleukin-18 Regulation of Interferon Gamma Production and Cell Proliferation as Shown in Interleukin-1beta-Converting Enzyme (Caspase-1)-Deficient Mice," *Blood* 91 (1998): 2118–2125.

47. Daniela Novick et al., "Interleukin-18 Binding Protein: A Novel Modulator of the Th1 Cytokine Response," *Immunity* 10 (1999): 127–136. I am extremely proud to be a co-author of this article.

48. Novick and Rubinstein, "The Tale."

49. Kakuji Torigoe et al., "Purification and Characterization of the Human Interleukin-18 Receptor," *J Biol Chem* 272 (1997): 25737–25742; Teresa L. Born et al., "Cloning of a Novel Receptor Subunit, AcPL, Required for Interleukin-18 Signaling," *J Biol Chem* 273 (1998): 29445–29450.

50. Novick et al., "Interleukin-18 Binding Protein." Our group had originally submitted the article to the even more prestigious journal *Cell*. The data in

the original publication included purification and cloning of IL-18BP as well as demonstration of its *in vitro* activity as an IL-18 decoy. "They had no real comments or corrections except the demand to perform an *in vivo* study," Daniela Novick reminded me. "Charles [Dinarello] suggested to treat endotoxin-injected mice with IL-18BP and measure interferon-γ in their blood." This suggestion was based on the properties of IL-18 as a major inducer of interferon-γ (reviewed in Chapter 3). In this experiment, endotoxin would induce production of IL-18, which would then upregulate synthesis of interferon-γ. Therefore, if IL-18BP indeed acted as an IL-18 decoy *in vivo* as it did *in vitro*, mice that received IL-18BP would have lower levels of interferon-γ compared to mice that had not been injected with IL-18BP. "I first calibrated endotoxin and only then used the recombinant IL-18BP that [Soo-Hyun] Kim had produced to treat mice," Dr. Novick continued. "It was such a responsibility that my hands trembled while injecting mice because, would the mice die or the IL-18BP dose not suffice, we were unable to repeat the experiment for many weeks. Fortunately everything worked. It was a hot August day and in addition to the pressure with this *in vivo* experiment there were many additional issues going on. We sent back the manuscript to *Cell* and their response was that the *in vivo* experiment we had chosen was not appropriate." The editors wanted a different type of experiment, specifically requesting the authors to test the effectiveness of IL-18BP in a model of parasitic infection induced by the protozoa *Leishmania* in mice. "We then found out why," Dr. Novick continued. "The editor in charge was a *Leishmania* expert. We could not make this model, so the paper was rejected by *Cell*."

51. For references to the original articles, see Charles A. Dinarello et al., "Interleukin-18 and IL-18 Binding Protein," *Front Immunol* 4 (2013): 289. For more information of the use of IL-18BP, see http://www.ab2bio.com/download /PressRelease_AB2_Bio_November2015_FINAL_EN.pdf as well as note 60 in Chapter 3.

52. Novick and Rubinstein, "The Tale." The "lunatic" quote is widely attributed to Australian politician Barry Jones.

53. There are many books and essays on viruses, presenting information from different perspectives and at all levels of sophistication. Though not an easy read, for a highly reasoned treatise on viruses, see *Viruses: Essential Agents of Life*, ed. Günther Witzany (Dordrecht, The Netherlands: Springer, 2012).

54. For references to original articles on viral cytokine decoys, see Megan L. Epperson, Chung A. Lee, and Daved H. Fremont, "Subversion of Cytokine Networks by Virally Encoded Decoy Receptors," *Immunol Rev* 250 (2012): 199–215.

CHAPTER 6   Between Inflammation and Metabolism:
The Acute-Phase Response

1. William S. Tillett and Thomas Francis, Jr., "Serological Reactions in Pneumonia With a Non-Protein Somatic Fraction of Pneumococcus," *J Exp Med* 52 (1930): 561–571; Jean Hurlimann, G. Jeannette Thorbecke and Gerald M. Hochwald, "The Liver as the Site of C-Reactive Protein Formation," *J Exp*

Med 123 (1966): 365–378. Also see the series of three articles related to bio-chemical characterization of CRP published by the group of Oswald T. Avery in *J Exp Med*, volume 73, 1941.

2. Production of certain liver proteins, called negative acute-phase proteins, actually decreases during inflammation. However, their quantitative decrease is not as dramatic as the increase in positive acute-phase proteins like CRP.

3. Freddy Homburger, "A Plasma Fibrinogen Increasing-Factor Obtained From Sterile Abscesses in Dogs," *J Clin Invest* 24 (1945): 43–45; Stephanie N. Vogel and Jean D. Sipe, "The Role of Macrophages in the Acute Phase SAA Response to Endotoxin," *Surv Immunol Res* 1 (1982): 235–241; David G. Ritchie and Gerald M. Fuller, "Hepatocyte-Stimulating Factor: A Monocyte-Derived Acute-Phase Regulatory Protein," *Ann NY Acad Sci* 408 (1983): 490–502; Aleksander Koj et al., "The Acute-Phase Response of Cultured Rat Hepato-cytes. System Characterization and Effect of Human Cytokines," *Biochem J* 224 (1984): 505–514; Heinz Baumann et al., "Human Keratinocytes and Monocytes Release Factors Which Regulate the Synthesis of Major Acute Phase Plasma Proteins in Hepatic Cells from Man, Rat, and Mouse," *J Biol Chem* 259 (1984): 7331–7342.

4. For very interesting, though rather technical, memoirs on the identification of IL-4, IL-5, and IL-6 and references to the original reports, see Eva Sever-inson, "Identification of the IgG1 Induction Factor (Interleukin 4)," *Front Immunol* 5 (2014): 628; William E. Paul, "History of Interleukin-4," *Cyto-kine* 75 (2015): 3–7; Kiyoshi Takatsu, "Revisiting the Identification and cDNA Cloning of T Cell-Replacing Factor/Interleukin-5," *Front Immunol* 5 (2014): 639; Toshio Hirano, "Revisiting the 1986 Molecular Cloning of Inter-leukin 6," *Front Immunol* 5 (2014): 1–456.

5. Jack Gauldie et al., "Interferon β_2 / B-Cell Stimulatory Factor Type 2 Shares Identity With Monocyte-Derived Hepatocyte-Stimulating Factor and Regu-lates the Major Acute Phase Protein Response in Liver Cells," *PNAS* 84 (1987): 7251–7255. As Dr. Rose-John pointed out in an e-mail about this dis-covery, "At the time, Dr. Peter Heinrich had heard about Jack Gauldie's finding at a meeting held at Cold Spring Harbor. Heinrich wrote a short article about this [Tilo Andus et al., "Recombinant Human B Cell Stimula-tory Factor 2 (BSF-2/IFN-β) Regulates β-Fibrinogen and Albumin mRNA Levels in Fao-9 Cells," *FEBS Letters* 221 (1987): 18–22], which was published before the paper by Jack Gauldie in *PNAS*. This generated a lot of bad blood between the two, since Gauldie felt scooped by Heinrich. Actually, the Gauldie *PNAS* paper had been submitted seven weeks before the Heinrich report and this latter did mention the talk of Jack Gauldie at the Cold Spring Harbor Meeting." IL-6 also turned out to overlap with a protein of unknown function induced by stimulation with IL-1 and TNF that had been named interferon β-2. Hybridomas are cells generated in the laboratory by fusing a B lymphocyte with a myeloma cell with the goal of producing antibodies that are called monoclonal because they are all identical and derive from a single clone of cells. Development of this technique, widely used in research and essential to generate therapeutic antibodies for the clinic, was at the origin of

the 1984 Nobel Prize in Physiology or Medicine to César Milstein and Georges J. F. Köhler.

6. Dr. Kishimoto's quotes and personal information are from a February 2014 e-mail interview.

7. From a February 2014 e-mail interview. For a memoir of these discoveries and references to the original reports, see Tadamitsu Kishimoto, "IL-6: From Its Discovery to Clinical Applications," *Int Immunol* 22 (2010): 347–352.

8. For references to the original reports, see Kishimoto, "IL-6.". For more on IL-6 and its signaling pathways, see Cristoph C. Garbers, Samadhi Aparicio-Siegmund, and Stefan Rose-John, "The IL-6/gp130/STAT3 Signaling Axis: Recent Advances Towards Specific Inhibition," *Curr Opin Immunol* 34 (2015): 75–82. For more on the JAK/STAT pathway, see Alejandro V. Villarino et al., "Mechanisms of Jak/STAT Signaling in Immunity and Disease," *J Immunol* 194 (2015): 21–27.

9. Remarkably, despite their very different biological outcomes, IL-6 and IL-10 signal by activation of the very same STAT-3/SOCS-3 pathway. Patterns of receptor expression on different cell types as well as sensitivity of the signaling pathways contribute to determine the type of biological outcome conveyed by IL-6 versus IL-10.

10. See Stefan Rose-John and Peter C. Heinrich, "Soluble Receptors for Cytokines and Growth Factors: Generation and Biological Function," *Biochem J* 300 (1994): 281–290.

11. Quotes and information by Dr. Rose-John are from an April 2014 e-mail interview.

12. Tetsuya Taga et al., "Interleukin-6 Triggers the Association of Its Receptor With a Possible Signal Transducer, gp130," *Cell* 58 (1989): 573–581.

13. Daniela Novick et al., "Soluble Cytokine Receptors are Present in Normal Human Urine," *J Exp Med* 170 (1989): 1409–1414.

14. Matsuo Honda et al., "Human Soluble IL-6 Receptor: Its Detection and Enhanced Release by HIV Infection," *J Immunol* 148 (1992): 2175–2180; Andrzej Machiewicz et al., "Soluble Human Interleukin-6-Receptor Modulates Interleukin-6-Dependent N-Glycosylation of Alpha-1 Protease Inhibitor Secreted by HepG2 Cells," *FEBS Lett* 306 (1992): 257–261; Daniela Novick et al., "Enhancement of Interleukin 6 Cytostatic Effect on Human Breast Carcinoma Cells by Soluble IL-6 Receptor from Urine and Reversion by Monoclonal Antibody," *Cytokine* 4 (1992): 6–11; Andrzej Machiewicz et al., "Soluble Interleukin 6 Receptor Is Biologically Active In Vivo," *Cytokine* 7 (1995): 142–149.

15. Jürgen Müllberg et al., "Protein Kinase C Activity Is Rate Limiting for Shedding of the Interleukin-6 Receptor," *Biochem Biophys Res Commun* 189 (1992): 794–800; Jürgen Müllberg et al., "The Soluble Interleukin-6 Receptor Is Generated by Shedding," *Eur J immunol* 23 (1993): 473–480. "We first submitted our paper on shedding of the IL-6 receptor to *EMBO Journal*," Dr. Rose-John wrote in an e-mail, referring to his submission to a journal considered more prestigious than the *European Journal of Immunology* that eventually accepted the article. "That journal rejected the paper with the comment, 'I am afraid that the paper requires considerably more experimental

data, particularly more control data to justify some of the conclusions presented, and even if the controls are added, I am not sure it would be suitable for publication in *EMBO Journal*.' In the end the article was published by the *European Journal of Immunology* in 1993 and has been cited until now 309 times and is frequently cited even 22 years after publication. The paper would have been good for the impact factor of *EMBO Journal*!" The highly cited review article is Rose-John, "Soluble Receptors."

16. These cells are called Th17 because they produce the cytokine IL-17. Th17 lymphocytes are the third type of T helper cells after Th1 and the Th2 types that were presented in Chapter 4.

17. Christopher A. Hunter and Simon A. Jones, "IL-6 as a Keystone Cytokine in Health and Disease," *Nat Immunol* 16 (2015): 448–457.

18. In Japan, anti IL-6 receptor antibodies are also approved for treatment of Castleman disease, a rare condition that looks like a blood cancer but is not cancer. For more information on current and prospected use of anti-IL-6 therapies, see Xin Yao et al., "Targeting Interleukin-6 in Inflammatory and Autoimmune Diseases and Cancer," *Pharmacol Therapeut* 141 (2014): 125–139.

19. It is important to note that not every scientist agrees that signaling mechanisms are the main factor that distinguishes the kind from the nasty sides of IL-6.

20. See, for example, Elena Fattori et al., "Defective Inflammatory Responses in Interleukin 6-Deficient Mice," *J Exp Med* 180 (1994): 1243–1250.

21. For information on the evolution of acute-phase proteins and other molecules involved in innate immunity, see Brad G. Magor and Kathy E. Magor, "Evolution of Effectors and Receptors of Innate Immunity," *Dev Comp Immunol* 25 (2001): 651–682.

22. There are several versions of pentraxins, some of which are directly involved in inflammation, as shown—among others—by Dr. Mantovani's group. See, for example, Antonio Inforzato et al., "Pentraxins in Humoral Innate Immunity," *Adv Exp Med Biol* 956 (2012): 1–20.

23. Other acute-phase proteins substitute for CRP in mice, a possible candidate being serum amyloid P, another member of the pentraxin family.

24. There are limits to the protective effects of human CRP in mice, as this intervention does not work under every experimental condition. For references to the original articles, see Gideon H. Hirschfield et al., "Human C-Reactive Protein Does Not Protect Against Acute Lipopolysaccharide Challenge in Mice," *J Immunol* 171 (2003): 6046–6051; Andres Peisajovish et al., "C-Reactive Protein at the Interface Between Immunity and Inflammation," *Expert Revi Clin Immunol* 4956 (2008): 379–390.

25. Steven Black, Irving Kushners, and David Samols, "C-Reactive Protein," *J Biol Chem* 279 (2004): 48487–48490; Nicholas R. Jones et al., "Collagen-Induced Arthritis Is Exacerbated in C-Reactive Protein-Deficient Mice," *Arthritis Rheum* 63 (2011): 2641–2650; J. Paul Simon et al., "C-Reactive Protein Is Essential for Innate Resistance to Pneumococcal Infection," *Immunology* 143(2014): 414–420.

26. For a discussion of these issues, see Michael Torzewski, Ahmed Bilal Waqr,

and Jianglin Fan, "Animal Models of C-Reactive Protein" *Mediat Inflamm* 2014 (2014): 683598.

27. For references to the original articles, see Frederick Strang and Heribert Schunkert, "C-Reactive Protein and Coronary Heart Disease: All Said—Is Not It?" *Mediat Inflamm* 2014 (2014): 757123.

28. Torzewski et al., "Animal Models."

29. Strang and Schunkert, "C-Reactive Protein." Associations of CRP genetic variants with risk of bacterial infections and autoimmune diseases have also been demonstrated, though the mechanisms involved are not completely clear.

30. Ibid. The mechanisms linking IL-6 receptor variants to risk of cardiovascular disease are under discussion. Other cytokines, including IL-1 and other inflammatory mediators, are involved in the pathogenesis of cardiovascular disease.

31. Quotations are from a March 2014 e-mail interview, in which Dr. Sipe also stated, "It was an exciting time [at the NIH]. Because of the Vietnam War, many medical doctors chose to fulfill their military service in the Public Health Service at the NIH. Space was at a premium, but intellectual stimulation and research supplies were abundantly available. In 1980 I moved to Boston University, where I continued my amyloid research until 1997, when I returned to the NIH as a scientific review officer and I greatly enjoyed managing the evaluation of multidisciplinary grant applications in the area of musculoskeletal tissue engineering. After retirement in 2012, I have served as a peer review consultant."

32. It was Dr. Glenner's work in the early 1980s (he had moved from the NIH to the University of California, San Diego, by then) that led to identification of β-amyloid precursor as the protein that forms aggregates in the brain of people with Alzheimer's disease.

33. The protein that forms fibril deposits in amyloidosis is actually a fragment of SAA, called AA. Like CRP, which does not behave in mice as it does in humans, there is a little quirk to SAA when it comes to experimental animals: rats do not make SAA-1 and SAA-2, though they do have SAA-4. Gunilla T. Westermark et al., "AA Amyloidosis: Pathogenesis and Targeted Therapy," *Annu Rev Pathol* 10 (2015): 321–344; Fernando Kempta Lekpa et al., "Les Amyloses en Afrique Subsaharienne," *Médecine et Santé Tropicales* 22 (2012): 275–278.

34. See, for example, Jean D. Sipe et al., "Detection of a Mediator Derived from Endotoxin-Stimulated Macrophages That Induces the Acute Phase Serum Amyloid A Response in Mice," *J Exp Med* 19 (1979): 597–606; Marcelo B. Sztein et al., "The Role of Macrophages in the Acute-Phase Response: SAA Inducer Is Closely Related to Lymphocyte Activating Factor and Endogenous Pyrogen," *Cell Immunol* 63 (1981): 164–176; Ruth Neta et al., "Comparison of In Vivo Effects of Human Recombinant IL 1 and Human Recombinant IL 6 in Mice," *Lymphokine Res* 7 (1988): 403–412. Many more of Dr. Sipe's articles examine the inflammatory determinants of SAA induction.

35. Kari K. Eklund et al., "Immune Functions of Serum Amyloid A," *Crit Rev Immunol* 32 (2012): 335–348; Driss El Kebir, Levente József, and János G.

Filep, "Opposing Regulation of Neutrophil Apoptosis Through the Formyl Peptide Receptor-like 1/Lipoxin A_4 Receptor: Implications for Resolution of Inflammation," *J Leukoc Biol* 84 (2008): 600–606.

36. Eklund et al., "Immune Functions."

37. Christina H. Park et al., "Hepcidin, a Urinary Antimicrobial Peptide Synthesized in the Liver," *J Biol Chem* 276 (2001): 7806–7810; Alexander Krause et al., "LEAP-1, a Novel Highly Disulfide-Bonded Human Peptide, Exhibits Antimicrobial Activity," *FEBS Lett* 480 (2000): 147–150; Christelle Pigeon et al., "A New Mouse Liver-specific Gene, Encoding a Protein Homologous to Human Antimicrobial Peptide Hepcidin, Is Overexpressed during Iron Overload," *J Biol Chem* 276 (2001): 7811–7819. For a detailed history of hepcidin's discovery, see Tomas Ganz, "Hepcidin and Iron Regulation, 10 Years Later," *Blood* 117 (2011): 4425–4433.

38. Gaël Nicolas et al., "Lack of Hepcidin Gene Expression and Severe Tissue Iron Overload in Upstream Stimulatory Factor 2 (USF2) Knockout Mice," *PNAS* 98 (2001): 8780–8785. Also see Ganz, "Hepcidin."

39. The inspiration for these comments is the wonderful story of the word *serendipity* by Robert King Merton and Elinor G. Barber, *The Travels and Adventures of Serendipity: A Study in Sociological Semantics and the Sociology of Science* (Princeton, NJ: Princeton University Press, 2004). The modern definition of serendipity is from *Collins English Dictionary—Complete and Unabridged* (1991, 1994, 1998, 2000, 2003). Retrieved June 21, 2015. I am certainly not the first to bemoan the loss of the sagacity part from the definition of serendipity. For example, see Morton A. Meyers, *Happy Accidents. Serendipity in Major Medical Breakthroughs in the Twentieth Century* (New York: Arcade Publishing, 2007).

40. Ganz, "Hepcidin."

41. Because enterocytes die and are replaced every few days, trapped iron leaves the body when these dead cells are shed in the feces.

42. The quote is from an April 2014 e-mail interview with Dr. Nemeth. Also see Elizabeta Nemeth et al., "Hepcidin Regulates Cellular Iron Efflux by Binding to Ferroportin and Inducing Its Internalization," *Science* 306 (2004): 2090–2093. Drs. Nemeth and Ganz also led the group that identified the hormone that shuts down hepcidin production after blood loss. See Léon Kautz et al., "Identification of Erythroferrone as an Erythroid Regulator of Iron Metabolism," *Nat Genet* 46 (2014): 678–684. Regulation of iron metabolism involves many more steps and components than I mentioned, with control not only of absorption but also of iron compartmentalization as well as interchanges among various molecular forms.

43. See references in Elizabeta Nemeth and Tomas Ganz, "Anemia of Inflammation," *Hematol Oncol Clin North Am* 28 (2014): 671–681.

44. Elizabeta Nemeth et al., "Hepcidin, a Putative Mediator of Anemia of Inflammation, Is a Type II Acute-Phase protein," *Blood* 101 (2003): 2461–2463; Elizabeta Nemeth et al., "IL-6 Mediates Hypoferremia of Inflammation by Inducing the Synthesis of the Iron-regulatory Hormone Hepcidin," *J Clin Invest* 113 (2004): 1271–1276. Depending on the specific disease or experimental conditions, mediators other than IL-6 induce hepcidin.

45. Günther Weiss and Georg Schett, "Aneamia in Inflammatory Rheumatic Diseases," *Nat Rev Rheum* 9 (2013): 205–215. Also see Nemeth and Ganz, "Anemia of Inflammation."

46. See, for example, Alexandros N. Vgontzas et al., "Elevation of Plasma Cytokines in Disorders of Excessive Sleepiness: Role of Sleep Disturbance and Obesity," *J Clin Endocrinol Metab* 82 (1997): 1313–1316; Marjolein Visser et al., "Elevated C-Reactive Protein Levels in Overweight and Obese Adults, *JAMA* 282 (1999): 2131–2135.

47. Margaret F. Gregor and Gokhan Hotamisligil, "Inflammatory Mechanisms in Obesity," *Annu Rev Immunol* 29 (2011): 415–445.

48. Hua Yu et al., "Revisiting STAT3 Signalling in Cancer: New and Unexpected Biological Functions," *Nat Rev Cancer* 14 (2014): 736–746; Jillian A. Fontes, Noel R. Rose, and Daniela Cihákova, "The Varying Faces of IL-6: From Cardiac Protection to Cardiac Failure," *Cytokine* 74 (2015): 62–68. The Fontes review discusses the ambivalent role of acute versus chronic elevations of IL-6 in the context of heart disease.

49. Diabetes figures are from http://www.who.int/mediacentre/factsheets /fs312/en/.

50. Hunter and Jones, "IL-6 as a Keystone"; Jan Mauer, Jesse L. Denson, and Jens C. Brüning, "Versatile Functions for IL-6 in Metabolism and Cancer," *Trends Immunol* 36 (2015): 92–101; Michael J. Kraakman et al., "Blocking IL-6 Trans-Signaling Prevents High-Fat Diet-Induced Adipose Tissue Macrophage Recruitment but Does Not Improve Insulin Resistance," *Cell Metab* 3 (2015): 403–416.

51. Lisa Tussing-Humphreys et al., "Rethinking Iron Regulation and Assessment in Iron Deficiency, Anemia of Chronic Disease, and Obesity: Introducing Hepcidin," *J Acad Nutri Diet* 112 (2012): 391–400.

52. David Bohm, quoted in William Byers, *How Mathematicians Think. Using Ambiguity, Contradiction, and Paradox to Create Mathematics* (Princeton, NJ: Princeton University Press, 2010), 25.

CHAPTER 7   The Fat Revolution: Adipokines

1. My use of "human race" does not—in any way—constitute an endorsement of the existence of biologically distinct human races. Among the many excellent discussions on the construct of race from a biological, anthropological, cultural, and social perspective, see Dorothy Roberts, *Fatal Invention: How Science, Politics and Big Business Re-Create Race in the Twenty-First Century* (New York: New Press, 2011), as well as the collection of articles and commentaries in the supplement Genetics for the Human Race, *Nature Genetics* 36 (2004): 11s.

2. For a history of the Hottentot Venus, see Sadiah Qureshi, "Displaying Sara Baartman, the Hottentot Venus," *Hist Sci* 42 (2004): 233–257. For details on the role of adipose tissue in generating anatomical differences between Asian and Caucasian upper eyelids, see Sangki Jeong et al., "The Asian Upper Eyelid: An Anatomical Study with the Comparison to the Caucasian Eyelid," *Arch Ophthalmol* 117 (1999): 907–912. For a report on the popularity

of plastic surgery—mostly discussing changing eye shape toward a more Caucasian-looking type—in South Korea, see Patricia Marx, "About Face. Why Is South Korea the World's Plastic-Surgery Capital?" *The New Yorker*, March 23, 2015.

3. For more information on the role of adipose tissue as an energy store and a defense against lipid toxicity, see, for example, Jean-Pierre Montmayeur and Johannes le Coutre, eds., "Part 1: Importance of Dietary Fat," *Fat Detection: Taste, Texture, and Post-Ingestive Effects.* (Boca Raton, FL: CRC Press, 2010); Roger H. Unger, Philipp E. Scherer, and William L. Holland, "Dichotomous Roles of Leptin and Adiponectin as Enforcers Against Lipotoxicity During Feast and Famine," *Mol Biol Cell* 24 (2013): 3011–3015.

4. For more information on adipose tissue, see Philip A. Wood, *How Fat Works* (Cambridge, MA: Harvard University Press, 2009). For technical information on adipose tissue and its functioning, see Carol A. Braunschweig and Giamila Fantuzzi, eds., *Adipose Tissue and Adipokines in Health and Disease*, 2nd ed. (Totowa, NJ: Humana Press Springer, 2014).

5. Yiying Zhang et al., "Positional Cloning of the Mouse Obese Gene and its Human Homologue," *Nature* 371 (1994): 425–432.

6. For a review of these early observations, see George A. Bray, "Commentary on Classics of Obesity. 4. Hypothalamic Obesity," *Obes Res* 1 (1993): 325–328.

7. See John R. Brobeck, "Mechanism of the Development of Obesity in Animals with Hypothalamic Lesions," *Physiol Rev* 26 (1946): 541–559, and references therein.

8. Gordon C. Kennedy, "The Role of Depot Fat in the Hypothalamic Control of Food Intake in the Rat," *Proc R Soc Lond B: Biol Sci* 140 (1953): 578–596.

9. Although the outcome of some of the parabiosis experiments I am about to discuss can be hard to tolerate, we need to remember that scientists take the outmost care in making sure animals do not unnecessarily suffer from laboratory procedures. This is not only an ethical requirement legally imposed by animal care committees, but also a necessity if one is to be able to interpret the results of an experiment correctly. An animal who is in pain or under severe distress behaves differently from an unstressed one and the researcher becomes unable to disentangle the effects of pain or stress from those of the specific question being tested. Parabiotic experiments, as well as other approaches that employ experimental animals discussed in this chapter and the preceding ones, including those that involve brain lesions, raise countless ethical questions concerning which level of animal suffering—if any—is acceptable in biomedical research. As already mentioned, these are complex ethical and moral issues that only an in-depth discussion can give justice to. I therefore refer the reader to the many excellent texts that address the ethics of animal use in research, including those referenced in Chapter 2, note 11.

10. G. Romaine Hervey, "The Effects of Lesions in the Hypothalamus in Parabiotic Rats," *J Physiol* 145 (1959): 336–352. Current ethical standards would not allow researchers to reach this end point, as animals that show signs of distress signaling impending death need to be humanely euthanized. In 2013,

at the age of eighty-eight, Romaine Hervey wrote a very interesting academic autobiography, G. Romaine Harvey, "Control of Appetite. Personal and Departmental Recollections," *Appetite* 61 (2013): 100–110.

11. The history of the Jackson Laboratory is from http://www.jax.org/fastfacts /index.html.

12. Ibid.

13. Ann M. Ingalls, Margaret M. Dickie, and George D. Snell, "Obese, a New Mutation in the House Mouse," *J Hered* 41 (1950): 317–318.

14. James B. Young and Lewis Landsberg, "Diminished Sympathetic Nervous System Activity in Genetically Obese (ob/ob) Mouse," *Am J Physiol* 245 (1983): E148–E154; Hiroyuki Kawashima and Albert Castro, "Effect of 1 Alpha-Hydroxyvitamin D3 on the Glucose and Calcium Metabolism in Genetic Obese Mice," *Res Comm Chem Path Pharmacol* 33 (1981): 155–161; Ranjit K. Chandra, "Cell-Mediated Immunity in Genetically Obese C57BL/6J ob/ob mice," *Am J Clin Nutr* 33 (1980): 13–16.

15. Katharine P. Hummel, Margaret M. Dickie, and Douglas L. Coleman, "Diabetes, a New Mutation in the Mouse," *Science* 153 (1966): 1127–1128.

16. For references and further discussion of the original articles, see Ruth Harris, "Contribution Made by Parabiosis to the Understanding of Energy Balance Regulation," *Biochim Biophys Acta* 1832 (2013): 1449–1455.

17. Douglas L. Coleman, "Effects of Parabiosis of Obese with Diabetes and Normal Mice," *Diabetologia* 9 (1973): 294–298.

18. Dr. Friedman's quote is from http:// laskerfoundation.org/media/v_coleman .htm.

19. Zhang et al., "Positional Cloning."

20. Mary Ann Pelleymounter et al., "Effects of the Obese Gene Product on Body Weight Regulation in ob/ob Mice," *Science* 269 (1995): 540–543; Jeffrey L. Halaas et al., "Weight-Reducing Effects of the Plasma Protein Encoded by the Obese Gene," *Science* 269 (1995): 543–546; L. Arthur Campfield et al., "Recombinant Mouse OB Protein: Evidence for a Peripheral Signal Linking Adiposity and Central Neural Networks," *Science* 269 (1995): 546–549. The Google search on the term *leptin* was performed on March 22, 2015.

21. Louis A. Tartaglia et al., "Identification and Expression Cloning of a Leptin Receptor, OB-R," *Cell* 83 (1995): 1263–1271; Hong Chen et al., "Evidence That the Diabetes Gene Encodes the Leptin Receptor: Identification of a Mutation in the Leptin Receptor Gene in db/db Mice," *Cell* 84 (1996): 491–495; Gwo-Hwa Lee et al., "Abnormal Splicing of the Leptin Receptor in Diabetic Mice," *Nature* 379 (1996): 632–645; Streamson C. Chua Jr. et al., "Phenotypes of Mouse Diabetes and Rat Fatty Due to Mutations in the OB (Leptin) Receptor," *Science* 271 (1996): 994–996; Michael Rosenbaum and Rudolph L. Leibel, "Role of Leptin in Energy Homeostasis in Humans," *J Endocrinol* 223 (2014): T83–T96. Dr. Daniela Novick, who had used affinity chromatography to isolate receptors for various cytokines, as we saw in previous chapters, told me how she attempted to use the same technique to identify the leptin receptor from urine, but failed to find it because this receptor was too large to pass through the kidney and end up in urine.

22. Dr. Myers's quotes and personal information are from an April 2014 e-mail interview.

23. For more on leptin's biology and discovery, see articles in "Leptin in the 21st Century," *Metabolism* 64 (2015): 1–156. Dr. Myers's quote are from an April 2014 e-mail interview.

24. Richard Stone, "Rockefeller Strikes Fat Deal With Amgen," *Science* 268 (1995): 631.

25. Wade Roush, "Fat Hormone Poses Hefty Problem for Journal Embargo," *Science* 269 (1995): 627.

26. Donald F. Phillips, "Leptin Passes Safety Tests, but Effectiveness Varies," *JAMA* 280 (1998): 869–870; Steven B. Heymsfield et al., "Recombinant Leptin for Weight Loss in Obese and Lean Adults: A Randomized, Controlled, Dose-Escalation Trial," *JAMA* 282 (1999): 1568–1575.

27. Carl T. Montague et al., "Congenital Leptin Deficiency Is Associated With Severe Early-Onset Obesity in Humans," *Nature* 387 (1997): 903–908; Andreas Strobel et al., "A Leptin Missense Mutation Associated With Hypogonadism and Morbid Obesity," *Nat Genet* 18 (1998): 213–215.

28. Dr. Clément's quotes and personal information are from a February 2014 e-mail interview.

29. Quotes and personal information are from a February 2014 e-mail interview with Dr. Clément.

30. Karine Clément et al., "A Mutation in the Human Leptin Receptor Gene Causes Obesity and Pituitary Dysfunction," *Nature* 392 (1998): 398–401. Also see Karine Clément, "Genetics of Human Obesity," *Proc Nutr Soc* 64 (2005): 133–142; Agatha A. van der Klaauw and I. Sadaf Farooqi, "The Hunger Genes: Pathways to Obesity," *Cell* 161 (2015): 119–132.

31. E-mail interview, February 2014. The phenotype of individuals with leptin receptor mutations is somewhat variable: although extreme obesity is a common trait, for reasons that are still unclear some affected individuals never reach puberty and remain infertile, whereas others—like Dr. Clément's patient—are able to conceive and carry pregnancies to term. Unfortunately, Dr. Clément's other patient, the sister, eventually succumbed to the consequences of her condition.

32. Fredrik Lönngvist et al., "Overexpression of the Obese (ob) Gene in Adipose Tissue of Human Obese Subjects," *Nat Med* 1 (1995): 950–953; Robert C. Frederich et al., "Leptin Levels Reflect Body Lipid Content in Mice: Evidence for Diet-Induced Resistance to Leptin Action," *Nat Med* 1 (1995): 1311–1314.

33. See Martin G. Myers, Jr. et al., "Challenges and Opportunities of Defining Clinical Leptin Resistance," *Cell Metab* 15 (2012): 150–156.

34. In 2006, Amylin Pharmaceuticals acquired the rights to leptin's commercialization from Amgen, and in 2012 Bristol-Myers Squibb acquired Amylin itself. As of this writing, metreleptin is commercialized through a partnership between Bristol-Myers Squibb and AstraZeneca (keeping track of companies' mergers and acquisitions can be harder than following body messages). For more information on lipodystrophy, see Nivedita Patni and Abhimanyu Garg, "Congenital Generalized Lipodystrophies—New Insights into Metabolic Dysfunction," *Nat Rev Endocrinol* 11 (2015): 522–534.

35. See *Metabolism* Special Issue (2015) and Stephanie E. Simonds et al., "Leptin Mediates the Increase in Blood Pressure Associated with Obesity," *Cell* 159 (2014): 1404–1416.

36. Dr. Scherer's quote and personal information are from an e-mail interview dated February 2014. Dr. Scherer continued telling me the story of his scientific career, "I felt that the opportunities as an assistant professor were significantly better in the United States than in Switzerland. More room for growth, scientifically a more stimulating overall environment, particularly in my field of metabolism. I was lucky to get into a very supportive group at Albert Einstein College of Medicine in New York City, which enabled me to develop programmatically prior to moving to the University of Texas Southwestern, where I currently hold the Gifford O. Touchstone, Jr. and Randolph G. Touchstone Distinguished Chair in Diabetes Research. Moving back to Europe has always been a possible option, but with our kids born in the United States and completing their education here, it becomes logistically increasingly challenging. Furthermore, we have ideal working conditions here at the University of Texas Southwestern Medical Center in Dallas with a highly integrated metabolism group."

37. Philipp E. Scherer et al., "A Novel Serum Protein Similar to C1q, Produced Exclusively in Adipocytes," *J Biol Chem* 270 (1995): 26746–26749; Erding Hu et al., "AdipoQ Is a Novel Adipose-Specific Gene Dysregulated in Obesity," *J Biol Chem* 271 (1996): 10697–10703; Kazuhisa Maeda et al., "cDNA Cloning and Expression of a Novel Adipose Specific Collagen-Like Factor, apM1 (Adipose Most Abundant Gene Transcript 1)," *Biochem Bioph Res Co* 221 (1996): 286–289; Yasuko Nakano et al., "Isolation and Characterization of GBP28, a Novel Gelatin-Binding Protein Purified From Human Plasma," *J Biochem* 120 (1996): 803–812.

38. Yukio Arita et al., "Paradoxical Decrease of an Adipose-Specific Protein, Adiponectin, in Obesity," *Biochem Bioph Res Co* 257 (1999): 79–83.

39. See, for example, Noriyuki Ouchi et al., "Novel Modulator for Endothelial Adhesion Molecules: Adipocyte-Derived Plasma Protein Adiponectin," *Circulation* 100 (1999): 2473–2476; Kikuko Hotta et al., "Plasma Concentrations of a Novel, Adipose-Specific Protein, Adiponectin, in Type 2 Diabetic Patients," *Arterioscl Throm Vas* 20 (2000): 1595–1599; Noriyuki Ouchi et al., "Adiponectin, an Adipocyte-Derived Plasma Protein, Inhibits Endothelial NF-kappaB Signaling Through a cAMP-Dependent Pathway" *Circulation* 102 (2000): 1296–1301.

40. Toshimasa Yamauchi et al., "The Fat-Derived Hormone Adiponectin Reverses Insulin Resistance Associated with Both Lipoatrophy and Obesity," *Nat Med* 7 (2001): 941–946; Anders H. Berg et al., "The Adipocyte-Secreted Protein Acrp30 Enhances Hepatic Insulin Action," *Nat Med* 7 (2001): 947–953.

41. For more details and references to the original papers, see Aslan T. Turer and Philipp E. Scherer, "Adiponectin: Mechanistic Insights and Clinical Applications," *Diabetologia* 55 (2012): 2319–2326. To be sure, there are some situations in which adiponectin levels do not change in the expected direction and where adiponectin seems to have paradoxical effects. For more information

on this, see Saeed Esmaili, Amin Xu, and Jacob George, "The Multifaceted and Controversial Immunometabolic Actions of Adiponectin," *Trends Endocrinol Metab* 25 (2014): 444.451; Giamila Fantuzzi, "Adiponectin in Inflammatory and Immune-Mediated Diseases," *Cytokine* 64 (2013): 1–10.

42. Recall that high levels of CRP predict elevated risk of heart attacks, as discussed in Chapter 6. Tobias Pischon et al., "Plasma Adiponectin Levels and Risk of Myocardial Infarction in Men," *JAMA* 291 (2004): 1730–1737.

43. Hanieh Yaghootkar et al., "Mendelian Randomization Studies Do Not Support a Causal Role for Reduced Circulating Adiponectin Levels in Insulin Resistance and Type 2 Diabetes, *Diabetes* 62 (2013): 3589–3598; Marjan Mansourian and Shaghayegh H. Javanmard, "Adiponectin Gene Polymorphisms and Susceptibility to Atherosclerosis: A Meta-Analysis," *J Res Med Sci* 18 (2013): 611–616.

44. Miki Okada-Iwabu et al., "A Small-Molecule AdipoR Agonist for Type 2 Diabetes and Short Life in Obesity," *Nature* 503 (2013): 493–399. Also, see Turer and Scherer, "Adiponectin."

45. Toshimasa Yamauchi et al., "Cloning of Adiponectin Receptors That Mediate Antidiabetic Metabolic Effects," *Nature* 423 (2003): 762–769; Hiroaki Tanabe et al., "Crystal Structure of the Human Adiponectin Receptors," *Nature* 520 (2015): 312–316.

46. Unger et al., "Dichotomous."

47. Ibid.

48. Leptin plays the same trick, so that the satiety and starvation hormones complement each other, leptin being protective during nutrient abundance, when it is elevated, adiponectin during nutrient deprivation, when its own levels go up.

49. Ja-Young Kim et al., "Obesity-Associated Improvements in Metabolic Profile through Expansion of Adipose Tissue," *J Clin Invest* 117 (2007): 2621–2637. In this study, the authors did not check whether the location of adipose tissue in mutant mice switched from belly to skin. See also Unger et al., "Dichotomous"; Joseph M. Rutkowski, Jennifer H. Stern, and Philipp E. Scherer, "The Cell Biology of Fat Expansion," *J Cell Biol* 208 (2015): 501–512.

50. Unger et al., "Dichotomous"; Motoharu Awazawa et al., "Adiponectin Enhances Insulin Sensitivity by Increasing Hepatic IRS-2 Expression via a Macrophage-Derived IL-6-Dependent Pathway," *Cell Metab* 13 (2011): 401–412.

51. This translation from the original German quote is reported in Thure von Uexküll, "Endosemiosis," *Semiotica* 96 (1993): 5–51. In "Some Aspects of a Medical Anthropology: Pathic Existence and Causality in Viktor von Weizsäcker," *Hist Psychiatr* 20 (2009): 360–376. Hartwig Wiedebach locates the original version to Viktor von Weizsäcker, *Soziale Krankheit und soziale Gesundung* (1930) in Gesammelte Schriften, eds. P. Achilles, D. Janz, M. Schrenk, and C. F. von Weizsäcker, (Frankfurt, Germany: Suhrkamp, 1986–2005), vol. 8, 94.

EPILOGUE

1. For more information, see Giamila Fantuzzi and Raffaella Faggioni, "Leptin in the Regulation of Immunity, Inflammation, and Hematopoiesis," *J Leukoc Biol* 68 (2000): 437–446; Carl Grunfeld, "Leptin and the Immunosuppression of Malnutrition," *J Clin Endocrinol Metab* 87 (2002): 3038–3039; Giamila Fantuzzi, "Adipose Tissue, Adipokines and Inflammation," *J Allergy Clin Immun* 115 (2005): 911–919; Fortunata Carbone, Claudia La Rocca, and Giuseppe Matarese, "Immunological Functions of Leptin and Adiponectin," *Biochimie* 94 (2012): 2082–2088; Priya Duggal et al., "A Mutation in the Leptin Receptor Is Associated with *Entamoeba histolytica* Infection in Children," *J Clin Invest* 121 (2011): 1191–1198.

2. For example, see Bruce Spiegelman and Gokhan S. Hotamisligil, "Through Thick and Thin: Wasting, Obesity, and TNF alpha," *Cell* 73 (1993): 625–627; Carl Grunfeld and Ken R. Feingold, "The Metabolic Effects of Tumor Necrosis Factor and Other Cytokines," *Biotherapy* 3 (1991): 143–158. Given the propensity of biomedical researchers to change the names of molecules, it is remarkable how TNF managed to maintain its ever more incongruous moniker over many decades.

3. For example, see Marianne Böni-Schnetzler and Marc Y Donath, "How Biologics Targeting the IL-1 System Are Being Considered for the Treatment of Type 2 Diabetes," *Brit J Clin Pharmaco* 76 (2013): 263–268; Thomas Mandrup-Poulsen, AIDA Study Group, "Interleukin 1 Antagonists for Diabetes," *Expert Opin Inv Drug* 22 (2013): 965–979; Christian Herder and Marc Y. Donath, "Interleukin-1 Receptor Antagonist: Friend or Foe to the Heart?" *Lancet Diabetes Endocrinol* 3 (2015): 228–229; Laura C. O'Brien et al., "Interleukin-18 as a Therapeutic Target in Acute Myocardial Infarction and Heart Failure," *Mol Med* 20 (2014): 221–229.

4. Moritz Haneklaus and Luke O'Neill, "NLRP3 at the Interface of Metabolism and Inflammation," *Immunol Rev* 265 (2015): 53–62; Yun-Hee Youm et al., "The Ketone Metabolite β-hydroxybutyrate Blocks NLRP3 Inflammasome-Mediated Inflammatory Disease," *Nat Med* 21 (2015): 263–269; Mihai G. Netea and Leo A. B. Joosten, "Inflammasome Inhibition: Putting Out the Fire," *Cell Metab* 21 (2015): 513–514; Elisabeth Kugelberg, "Starving Inflammation," *Nat Rev Immunol* 15 (2015): 199.

Acknowledgments

There are many people I need to thank for this project, but I would like to acknowledge first the place where this book was born. In the winter of 2013–2014, I spent a month in Cayman Brac, a tiny, gorgeous island in the Caribbean Sea: my partner and I, two bicycles, a modest rental near the water, the only luxury the opportunity to let body and mind roam freely. I had brought along a few readings, including an issue of *Nature*. Toward the back of that magazine was a short article by Sarah Webb, *Popular Science: Get the Word Out*, which discussed the pleasures and difficulties of writing science books. I would probably not have read Webb's article had I not been in that quiet island, and even if I did, I would likely have forgotten it the next day, lost in my daily routine. But reading about the art of writing science books in that setting sparked an initially vague idea, which blossomed and flourished into this book. So, thank you Cayman Brac and thank you Sarah Webb.

A heartfelt acknowledgment to Harvard University Press editors Michael Fisher, who surprised me by accepting my proposal a few months before he retired, and Janice Audet, whose thoughtful and efficient advice has been invaluable as I was churning out one chapter after another. And many thanks to the whole staff at Harvard University Press for their helpful professionalism. My genuine appreciation also goes to Dr. Hannah Landecker for framing the book's content with her superb foreword.

Many individuals at my workplace, the University of Illinois at Chicago, deserve recognition, beginning with my department heads, Charles Walter and Ross Arena, who promptly granted me time to work on this book, and continuing with Katie Philippe, Emily Jordan, Randal Stone, and the rest of the staff of the Department of Kinesiology and Nutrition, whose efficient handling of administrative issues allowed me to focus on research and writing. A special thank goes to Amy McNeil—the very first reader of my very first pages—who encouraged me to pursue the project and offered support and advice. I am also indebted to all my faculty colleagues, particularly Drs. Carol Braunschweig, Alan Diamond, Jake Haus, Tim Koh, Melinda Stolley, and Lisa Tussing-Humphreys for many enlightening discussions. Thank you to all the students, fellows, and trainees who have attended my classes and conducted research in the laboratory: it's only through interactions with each of you that I managed to reach at least some level of clarity in my writing. My boundless gratitude goes to Kevin O'Brien and the whole staff of the Health Sciences Library of the University of Illinois at Chicago, who made this work possible by fulfilling my repeated requests for reading material in an efficient and professional manner. Finally, a warm hug to all my UIC United Faculty colleagues and to the many friends and supporters who helped us build our labor union. An unintended but thoroughly welcome side effect of our seemingly endless meetings and deliberations has been the exposure of my science-trained mind to your many different ways of thinking and approaching problems, which inspired reflections that are now invisibly embedded in these pages.

I am grateful to my scientific mentors, Charles Dinarello and Pietro Ghezzi (who was too humble to accept to be interviewed for this book) for teaching me the art of science while contributing in countless additional ways to my personal growth. My keenest appreciation also goes to each of the researchers who agreed to be interviewed for the book and provided excellent feedback during the revision process: Drs. William Arend, Tamas Bartfai, Karine Clément, Jean-Michel Dayer, Charles Dinarello, Roderick Flower, Tadamitsu Kishimoto, Alberto Mantovani, Martin Myers, Jr., Elizabeta Nemeth, Daniela Novick, Joost Oppenheim, Mauro Perretti, Maria-Grazia Roncarolo, Stefan Rose-John, Philipp Scherer, Jean Sipe, Hiroko Tsutsui, and David Wallach. Even though I was unable to include his story, I nevertheless wish to express my gratitude to Dr. Jan Vilcek for his willingness to participate.

This book is dedicated to four very special human beings without whom I would not be the person I am. My parents Rosanna and Giulio, who always guided by example, never imposing limits or arbitrary decisions, never afraid or ashamed of their passions and convictions. My brother, Davide, my very first friend and teacher. My partner Sergio, who—just as these pages were seeing the light of day—sadly had to return to the universal cycle of the elements after feeding my body and my mind with boundless love for over a quarter century.

Grazie a tutti.

Index